T0259685

SpringerBriefs in Applied Sciences and Technology

SpringerBriefs present concise summaries of cutting-edge research and practical applications across a wide spectrum of fields. Featuring compact volumes of 50–125 pages, the series covers a range of content from professional to academic.

Typical publications can be:

- A timely report of state-of-the art methods
- An introduction to or a manual for the application of mathematical or computer techniques
- A bridge between new research results, as published in journal articles
- A snapshot of a hot or emerging topic
- An in-depth case study
- A presentation of core concepts that students must understand in order to make independent contributions

SpringerBriefs are characterized by fast, global electronic dissemination, standard publishing contracts, standardized manuscript preparation and formatting guidelines, and expedited production schedules.

On the one hand, **SpringerBriefs in Applied Sciences and Technology** are devoted to the publication of fundamentals and applications within the different classical engineering disciplines as well as in interdisciplinary fields that recently emerged between these areas. On the other hand, as the boundary separating fundamental research and applied technology is more and more dissolving, this series is particularly open to trans-disciplinary topics between fundamental science and engineering.

Indexed by EI-Compendex, SCOPUS and Springerlink.

More information about this series at http://www.springer.com/series/8884

Shaharin Anwar Sulaiman
Editor

Rotating Machineries

Aspects of Operation and Maintenance

 Springer

Editor
Shaharin Anwar Sulaiman
Department of Mechanical Engineering
Universiti Teknologi Petronas
Seri Iskandar, Perak, Malaysia

ISSN 2191-530X   ISSN 2191-5318   (electronic)
SpringerBriefs in Applied Sciences and Technology
ISBN 978-981-13-2356-0   ISBN 978-981-13-2357-7   (eBook)
https://doi.org/10.1007/978-981-13-2357-7

Library of Congress Control Number: 2018952874

This Springer imprint is published by the registered company Springer Nature Singapore Pte Ltd.
The registered company address is: 152 Beach Road, #21-01/04 Gateway East, Singapore 189721, Singapore

Preface

Rotating machineries, often referred to as turbomachines, are machines that transmit energy between a rotor and a fluid. They include turbines, compressors, and pumps and are widely used in industries, namely power generation, oil and gas production, and petrochemical. This book discusses the maintenance aspect of rotating machines, through the collection of various works from different authors. The book has been written to provide views of experienced engineers in the aspect of maintenance of rotating machines. The book will be useful as a reference for practicing engineers in the related industries mentioned earlier. The book will provide them with a glimpse of some of the problems associated with rotating machines or equipment in the field and help them to achieve maximum performance efficiency and high availability. The editor would like to express his gratitude to all the contributing authors for their time, effort, and dedication in preparing the manuscripts for the book.

Seri Iskandar, Malaysia   Shaharin Anwar Sulaiman

Contents

Contributors

Mohd Amin Abd Majid Universiti Teknologi PETRONAS, Perak, Malaysia

M. B. Baharom Universiti Teknologi PETRONAS, Seri Iskandar, Perak, Malaysia

A. T. Baheta Universiti Teknologi PETRONAS, Seri Iskandar, Perak, Malaysia

A. D. Fentaye Universiti Teknologi PETRONAS, Seri Iskandar, Perak, Malaysia

Suhaimi Hassan Universiti Teknologi PETRONAS, Seri Iskandar, Perak, Malaysia

Ismady Ismail TNB Repair and Maintenance, Selangor, Malaysia

K. P. Leong Universiti Teknologi PETRONAS, Seri Iskandar, Perak, Malaysia

Ainul Akmar Mokhtar Universiti Teknologi PETRONAS, Perak, Malaysia

Masdi Muhammad Universiti Teknologi PETRONAS, Perak, Malaysia

Freselam Mulubran Universiti Teknologi PETRONAS, Perak, Malaysia

Sundralingam Muthanandan PETRONAS, Kuala Lumpur, Malaysia

Mohammad Shakir Nasif Universiti Teknologi PETRONAS, Perak, Malaysia

M. Azrul Nizam B. M. Zahir PETRONAS Chemicals MTBE, Pahang, Malaysia

Khairul Anwar B. M. Nor PETRONAS, Kuala Lumpur, Malaysia

Shaharin Anwar Sulaiman Universiti Teknologi PETRONAS, Seri Iskandar, Perak, Malaysia

Hamdan Ya Universiti Teknologi PETRONAS, Seri Iskandar, Perak, Malaysia

Condition Monitoring and Assessment for Rotating Machinery

Sundralingam Muthanandan and Khairul Anwar B. M. Nor

Initially, when rotating machinery was used in industry, the maintenance practice was run to failure basis. This means that an equipment was being operated until it was unable to continue its intended function. This was the situation when corrective maintenance was later being introduced and implemented. As the technology progress and the rotating equipment becomes more critical to operation and business, preventive maintenance was introduced. The preventive maintenance task is usually recommended by the Original Equipment Manufacturer (OEM). This maintenance requires a fixed time base or operational hour-based intervention so that some of the critical components at a higher probability of failure can be inspected and replaced if required.

As time progress, rotating machinery becomes more critical to operation and more complex machine is being introduced into the rotating equipment. This made the reliability of the equipment becomes very crucial, equipment reliability started to be tracked, and unplanned downtime to machines became unacceptable. As such, condition monitoring technologies were evolved to assess the equipment health while in operation in order to understand and detect any incipient and hidden failures. These technologies enable some failure modes detectable early, and consequently, timely maintenance can be executed to prevent costly corrective maintenance. This technology, which is used to detect and predict failures, is also known as predictive maintenance. Thus, most of the oil and gas industry limits the corrective maintenance to non-critical equipment where the cost of corrective maintenance is more economical and justifiable as compared to preventive and predictive maintenance.

S. Muthanandan (✉) · K. A. B. M. Nor
PETRONAS, Kuala Lumpur, Malaysia
e-mail: sundra@petronas.com

© The Author(s), under exclusive licence to Springer Nature Singapore Pte Ltd. 2019
S. A. Sulaiman (ed.), *Rotating Machineries*, SpringerBriefs in Applied Sciences and Technology, https://doi.org/10.1007/978-981-13-2357-7_1

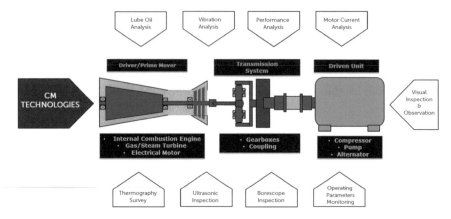

Fig. 1 Condition monitoring (CM) technologies for rotating machineries

Introduction

Condition monitoring is the use of special devices to measure specific aspects of equipment deterioration in order to diagnose the cause and recommend remedial action. Condition monitoring involves technical tasks in which operating parameters are periodically or continuously measured and recorded. The data will be later analyzed by comparing and displaying the data so that it could support any decision making for operation and maintenance of the rotating machinery.

Most of the industry practitioners and experts realize that unplanned downtime is very costly and time-consuming. Failure detrimental to equipment integrity could lead to organizational reputation and opportunity for value leakages. Therefore, detecting developing insipient failures or issues before they lead to catastrophic failures is one of the most challenging and imminent tasks. This detection not only provides insight into the operating condition but also provides direction to maintenance activities so that one can prevent unplanned downtime. With the development of various types of condition monitoring techniques including advance techniques, the detection and diagnosis becomes easier, enabling appropriate prescriptive actions to be taken in a timely manner.

Condition monitoring technologies for rotating machineries in plants, as shown in Fig. 1, mainly include the followings:

- Visual inspection and observation
- Vibration analysis
- Lube oil analysis
- Thermography—temperature distribution and contour analysis
- Ultrasonic inspection
- Motor current signature analysis
- Performance monitoring
- Equipment online operating parameters analysis

The condition monitoring process briefly comprises the following activities:

- Measurement, recording, and analysis of rotating equipment parameters to gauge rotating equipment health status.
- The best reference data for condition monitoring program would be the baseline data where the equipment is at its best mechanical condition. Subsequently, the existing condition or health of rotating equipment is then compared with baseline data, and analysis is performed to determine if there are any incipient failures or problems which are showing signs or patterns of degradation.

The technique in condition monitoring requires systematic measuring and trending of equipment parameters to assess the present conditions of rotating equipment, comparing against previous conditions and predicting the equipment remnant life. By doing so, remedial actions are taken only on as-needed basis. Condition monitoring identifies problems to equipment early, enabling failure preventive action to be taken before it escalates to a critical problem. It also allows the maintenance team to plan for spare parts, manpower, and optimized operational window to schedule the maintenance activities. Some of the key advantages of condition monitoring are:

  i. Detection of health of machine while in operation
 ii. Maintenance is only performed when required
iii. Significantly reduce major breakdowns
 iv. Increase equipment reliability
  v. Elimination of unnecessary costly time-based preventive maintenance
 vi. Non-stock parts can be ordered in advance
vii. Detection of maintenance-induced problems, especially when it involves major overhaul.

Visual Inspection/Observation

Visual inspection is one of the most economic and basic means of condition monitoring. This is one of the earliest condition monitoring techniques employed by the technicians using multi-sensory monitoring and adjustment, such as color coding, labeling, and descriptive criteria based on appearance, sound, touch, or smell. This technique normally involves daily walkabouts by the operators at the equipment to observe for signs of deterioration. Some of the observations are critical such as mechanical seal leaks for centrifugal pumps, tubing leaks, external corrosion, and coating deterioration. In addition, the operator's visual inspection covers lube oil discoloration, lube oil foaming, mechanical seal buffer or barrier tank oil discoloration, casing/metal surface hot spot, and abnormal sound. These are very crucial machine information which cannot be substituted by any other online data.

The expected machine information from the operator daily walkabout checks should include the following:

- Smell (odor issues, overheating equipment)
- Noise (abnormal noises such as knocking sound from reciprocating compressors, rough sound generated from the fin fan cooler bearings, hissing sound from pressurized hot air leak)
- Visual (exhaust emissions for internal combustion engine such as blue or white smoke)
- Touch with suitable glove or personal protective equipment (PPE), for detection of vibrations.

A structured round does not require all equipment to be checked in detail every shift. It is assumed that the frequency of checks will be based on equipment criticality to ensure an even spread of work that allows an operator to perform consistently a quality structured round.

A sample of visual observation checklist for gas engine-driven centrifugal pump is shown in Table 1. A good checklist should have the safe operating envelope which specifies the operating limits. This provides alert to the operators in case the reading is out of tolerance.

These data can be reviewed with the supervisor in charge and can be used for trending of any changes which could give an early warning sign of degradation.

Vibration Analysis

Vibration measurement is one of the best methods available for detecting and monitoring mechanical problems and health condition of rotating equipment. It is normal for Rotating Equipment (RE) to vibrate during operation. Rotating equipment that is operating in the best condition will have minimal vibration and noise levels. When RE vibration and noise increase, they imply incipient failures and mechanical fault, or sometimes process-induced failure. The root causes of the RE mechanical fault can be easily identified through its spectrum pattern, frequency dominant peak/amplitude (severity levels), and absolute/relative phase angle (direction of vibration).

Vibration is the movement of the casing and/or rotating component around a reference point. Vibration can be caused by numerous sources from design flaws, installation or assembly errors, manufacturing defect and maintenance or process/operation influences. These forces increase vibration levels due to unbalance, misalignment, bearing defects, aerodynamic forces, gear problems, and many others. Vibration data can be acquired and analyzed in the following forms:

  i. Increase or significant reduction of overall vibration value over time.
 ii. Transformation of the overall vibration signal/data into various sub-elements such as spectrum analysis using fast Fourier transform (FFT) technique.
iii. Utilization of basic vibration signal (waveform collected from X-Y proximity probes) to produce orbit plot, shaft center line plot, etc.

Table 1 Sample of site visual inspection/observation checklist for gas engine-driven pump

Gas engine	Low limit	High limit	Eng A	Eng B
Lube oil pressure				
Lube oil level				
Lube oil top up quantity				
Speed (RPM)				
Exhaust temperature				
Engine lube oil temperature				
Fuel pressure				
Engine jacket water temperature				
Visual inspection	Observations			
Governor linkages condition				
Spark plug and cable condition				
Engine skid mounting bolts condition				
Engine skid perimeter condition				
Sign of oil leak				
Lube oil color				
Any sign of cooling water leak				
Exhaust smoke discharge condition				
Exhaust pipe insulation condition				
Breather hose condition				
Engine sound				
Exhaust smoke condition				
Engine body condition				
Sign of oil leak at turbo charger				
Pump	Low limit	High limit	Pump A	Pump B
NDE bearing oil level				
DE bearing oil level				
Top up oil (consumption)				
Suction pressure				
Discharge pressure				
Pump discharge flow				
Visual inspection	Observations			
Any sign of oil or product leak				
Lube oil color				
NDE mechanical seal condition				
DE mechanical seal condition				

iv. Utilization of basic vibration signal (waveform and spectrum collected from *X*-*Y* proximity probes with keyphasor/tachometer) to produce transient plots such as Bode plot, waterfall plot, and others.

The primary purpose of vibration measurement is to collect, store, and detect the deviation from their baseline and compare with the safe operating limits (SOL).

Techniques of Vibration Analysis

With the advancements of today, digital analyzers are used to acquire data and analysis. This has enabled industries to utilize handheld data collectors and spectrum (FFT) analyzer, thus making data collection easier, more reliable, safer and provides higher levels of accuracy. Vibration magnitude is also known as "amplitude." This can be measured in terms of displacement, velocity, or acceleration.

Displacement

This measures the change of distance or position of an object relative to a reference point. On machinery, it is usually expressed in the unit of mils or micrometers (μm), and it is usually measured in peak-to-peak form. Proximity probes are used to directly measure relative displacement of the shaft. The displacement transducer (probe) detects proximity of conducting materials. The proximitor produces an eddy current in the coil, and the overall level of the voltage is determined by the cable and probe circuit impedance. The impedance of the circuit is varied by the distance of the conducting material.

The dual plane arrangement shown in Fig. 2 is necessary to determine the average shaft radial position. The displacement arrangement can also be used for position measurement such as axial thrust position and as a pulse for speed (keyphasor).

Velocity

Velocity is the rate of change of displacement. It measures the rate of the displacement change. Typical units for velocity are inches per second (in/s) or millimeters per second (mm/s) and zero-to-peak. It is measured with a velocity transducer. Velocity transducers have a mass such as a permanent magnet which in turn is supported by springs. The springs and magnet are surrounded by a damping fluid and an electrical coil. In some applications, the velocity probe with integrated piezo element is used. The vibration data which is initially measured in acceleration is converted into velocity by mathematical integration.

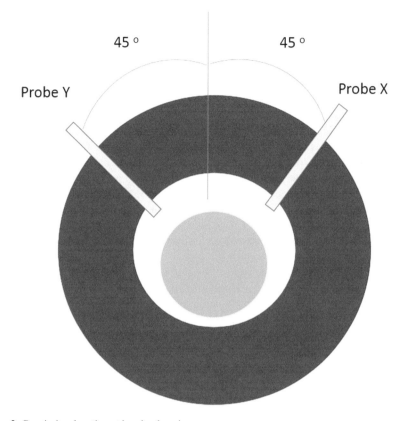

Fig. 2 Proximitor location at bearing housing

Acceleration

Acceleration is the rate of change of velocity. It measures the total force required to move the vibrating element in the opposite direction. Typical units for acceleration are inches per second per second (in/s^2) or millimeters per second per second (mm/s^2). In some cases, it is measured as gravitational acceleration, g which is usually given as $9.81 \ m/s^2$. The accelerometer is a piezoelectric transducer with charge amplifier, which generates an electrical signal that is proportional to acceleration.

Vibration Value

To enable correct trending of data, the way in which the vibration is measured has to be consistent. There are three ways of measuring the value. The ISO convention is to measure in RMS (root mean squared); however, it can also be measured in "peak to peak" (pk-pk) or "peak."

Trending

The overall vibration trending either via continuous online or periodic offline data will show indications of potential mechanical fault. The change seen in the trending shows mechanical incipient or degraded failures. So, the overall value trending will indicate the rate of deterioration so that operators could make necessary planning for rectification. As a practical approach, a baseline reading for vibration is taken and values are representative of the in situ condition when the machine is at its best mechanical condition. As such, the baseline reading is taken after a successful commissioning of the equipment or after a major overhaul. Thereafter, individual baseline values provide the reference against the monitored parameters during operation. This enables more reliable condition assessment to detect any changes to the machine.

Besides comparing against the baseline reading, the vibration readings are also compared against safe operating envelope, namely the warning or alert zone and the unsafe zone. Operating at the alert zone is allowed with increased level of monitoring in place. The risk of operation beyond the alert zone is mitigated via operators' frequent monitoring and increase in the level of vibration data acquisition and analysis. The operation at this region is not meant for long period but only as an interim operation until the equipment can be safely shut down for remedial action. Operating at an unsafe level is prohibited as this could lead to detrimental damages to equipment. If the equipment reaches this point, it is recommended to bring the equipment for a safe shutdown and execute maintenance. The safe operating envelope is determined based on the vibration standard or OEM limits for safe operation.

During the factory acceptance testing (FAT), the vibration limit such as overall and discrete peak shall be referenced to the user specification. Examples are User Technical Standards, if specified by the user the reference can be made to API standards such as API 610 for Centrifugal Pump, API 618 for Reciprocating Compressor and API 617 for Axial and Centrifugal Compressor and Expander–Compressor. The International Standard Organization (ISO) standards such as ISO 7919 Series and ISO 10816 Series also provide the vibration limits and safe operating envelopes which are normally used as a reference during operation. A sample of trending parameters and safe operating limit samples is shown in Fig. 3.

However, the overall vibration data and trending will not tell the causes of the high vibration. So, in order to perform troubleshooting, the data shall be transformed into other forms such as spectrum analysis in order to know the cause, and thereafter, to enable appropriate remedial action to be taken.

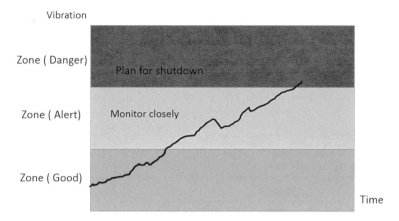

Fig. 3 Sample of trending on vibration and safe operating envelopes

Spectrum Analysis

Spectrum is a graph that shows vibration magnitude (amplitude) against frequency. The graph is produced as a result of transforming a signal from the time domain to the frequency domain. The composition of a time signal is transformed into a collection of sine waves. The procedure of doing the transformation is most commonly done with an FFT analyzer.

The vibration spectrum analysis enables analyses of vibration amplitudes at various component frequencies. By doing this, one can identify and track the vibration occurring at specific frequencies. This is because rotating equipment problems develop vibration peaks at specific frequencies, and this information is useful to diagnose the contributory causes. These components will be reference to equipment operating speed such as $1\times, 2\times$ running speed, gear mesh or impeller blade-pass frequencies and sidebands, rolling element bearing frequencies, and sub-synchronous energy associated with journal bearing instability.

Having variable duty machines, it is often found that vibration parameters are sensitive to other operating conditions such as operating speed and/or load. Hence, an increase in vibration amplitude is not necessarily indicative of deteriorating condition. As such, these changes should not be flagged as a problem. So, the analysis should not be concluded by merely looking into the vibration peaks and trends, but shall be associated with other operating parameters before concluding it as a mechanical problem.

Another essential function of the spectrum analysis is to understand the severity of the problem. For instance, high-frequency reading will normally indicate anti-friction bearing problem. So, by making reference to inspected bearing condition during maintenance and to the level of high-frequency reading, one can establish

Table 2 Vibration spectrum and equipment problem

No.	Problem	Vibration frequency	Causes
1.	Electrical problems	Synchronous AC line frequency	Common electrical problem such as broken rotor bars, eccentric rotor, unbalanced phases in multi-phase systems, and unequal air gap
2.	Oil whirl or oil whip or worn journal bearings, coupling damages, and resonance	$0.4 \times -0.48 \times$ operating Speed	Low stability at bearings, excessive bearing clearance, sub-harmonic resonance, improper oil properties
3.	Imbalance, lateral critical vibration, resonance, and shaft bow	$1 \times$ operating speed	Operating speed is at critical speed, casing distortion, rotor imbalance, coupling resonance and shaft or rotor bow
4.	Misalignment, mechanical looseness, bad belts ($2 \times$ RPM) and resonance	$2 \times$ operating speed	Reciprocating forces, low stability at bearings, excessive bearing and seal clearance, improper oil properties or worn bearings and seals
5.	Blade-pass frequency, caused by aerodynamic and hydraulic forces, and gear problem	Many times RPM, normally harmonic-related frequency (no. of blades $\times$ operating speed)	Operating speed is at critical speed, casing distortion, low flow in centrifugal pumps and short distance between centrifugal pump impeller and volute, reciprocating forces, mechanical looseness or bad gears
7.	Bearing problems, especially anti-friction bearings, gear damages, and shaft rubs	Very high frequency	Anti-friction bearing damages, support resonance, gear damages and internal rubbings, cavitation, recirculation and flow turbulence and improper lubrication to journal bearings

the severity of the damage. So, the maintenance inspection findings' reference to spectrum analysis can be further developed to establish specific damage severity. This would be helpful in the prediction of failure and planning for early intervention. The common issues that can be detected by spectrum analysis are listed in Table 2.

Lube Oil Analysis

Every rotating machine is unique in its operating environment because of its installed location, operating condition, fuels, and products encountered. The most critical aspect for accurate result of lube oil testing should be through the application of correct oil sampling and handling procedure. A specific plan shall be developed based on Reliability-Centered Maintenance (RCM) or Failure Mode Effect Analysis (FMEA) for actual frequency of sampling and types of testing based on the probability of failure and severity. Maintenance plans for lube oil include sampling of the oil, preservation, flushing, and also change out. The following subsections provide the general guidelines on the machinery lube oil and condition monitoring strategy. In addition to the lube oil testing, the lube oil consumption shall also be monitored especially for gas turbine and reciprocating engine.

Categories of Testing

Three main testing methods are:

i. **Visual Inspection** method, which is a critical preliminary indicator of equipment condition. This is done by allowing oil samples with settling time and later to perform visual inspection.
ii. **On-site Testing**, which uses the sample kits from lube oil testing laboratories. Having this instrument allows simple site testing where some critical testing such as water content, viscosity, and Total Acid Number (TAN) are tested and results can be obtained immediately.
iii. **Laboratory Testing**, which gives more accurate and advanced testing on the oil sample. This requires the oil samples to be transported to the laboratories, where more sophisticated testing is applied to provide detail information besides the visual and site testing.

Visual Inspection

Through visual inspection, a sample is taken from the lube system, and the sample is allowed to settle undisturbed for appropriate settling time.

- If the lube oil is found cloudy with a single layer, it is an indication of foaming. It can be caused by contamination or deficiency of anti-foaming agents in the lube oil.
- If the lube oil is found cloudy with a clear divided layer, it is an indication of water contamination. This is normally contributed by failures such as following:

- Seal leakage (especially steam turbine which eventually enters the bearing chamber and contaminate the oil)
- Oil cooler leakage
- Long-standby (e.g., firewater engine where the water leaks out from the pump packing and entered the bearing chamber)
- Clogged oil/mist eliminator system. This is generally applicable to the gas turbine lube oil system designed based on API 614. When the oil/mist eliminator is clogged, the mist is not removed from the lube oil system and thus remains in the oil.

- If the lube oil appears dark or blackish, it is an indication that the lube oil has been oxidized. The severity of oxidation depends on how dark the oil color is. Bearing operation at high temperature or occurrence of bearing fire could be one of the contributors.

On-Site Testing

Site testing requires portable kits which can be used to test the oil sample at site. Some of the testings are as follows;

- Water content
- Acid content or TAN
- Viscosity
- Solid contaminants (particle counts)

Findings from the site testing can be used as basis for further testing at the laboratory before a critical decision is made.

Lube Oil Laboratory Testing

Laboratory testing should be applied for samples found deteriorated or contaminated from on-site testing for detail analysis. This is to optimize the number of testing samples and the cost associated with the laboratory testing. The various types of analyses are listed in Table 3, in which different types of oil testing are provided. In addition to the lube oil analysis, it is important to monitor the lube oil consumption rate as part of routine operator's visual observation and records.

Table 3 Oil testing and purpose

No.	Oil testing	Purpose
1.	Viscosity testing	Crucial for retaining oil film thickness, too thick slow oil flow ability, too thin accelerate wear
2.	Flash point	It tells the tendency of an oil to form a flammable mixture with air
3.	Water content (Karl Fisher titration)	Water contamination badly affects the lubricant by acting as a catalyst for oxidation and depleting antioxidants. Water promotes corrosion and formation of varnish/sludge, which then leads to filter clogging
4.	Total acid number	Due to aging, acid by-products are formed. Strong acids are corrosive, while the weak acids attack bearing surfaces
5.	Oxidation stability test (RPVOT and RULER)	Measures the remaining antioxidants, thus the remaining useful life of the oil. Low values is contribute by the formation of free radicals in oil and this can lead to formation of varnish and sludge
6.	Spectrography	Identify wear components by chemical composition. It shows the start of abnormal wear and the components which are wearing. It assists in determining the root cause of wear/failure
7.	Ferrography	Identify metal particles, size, and origin. Oil with high % of large particles is abrasive to bearings and clogs filters/strainers

Note The testing method shall be referred to ASTM. The limits for the test result shall be referred to original equipment manufacturer (OEM)

Lube Oil Consumption Monitoring

The lube oil consumption rate tells the equipment health for some internal degradation. For example, high lube oil consumption for reciprocating engine may indicate worn out piston ring resulting to the lube oil leaks into combustion chamber. It may also mean a leaking gasket between the cylinder block and cylinder head. If this is noticed, other tests like compression test can be performed at the cylinders to confirm if there are any issues with piston rings or gaskets. This can be even further verified with the crankcase blow-by flow for any abnormalities. The exhaust gas with white smoke may indicate for burning oil, and this is an additional clue in the diagnostic.

High lube oil consumption on gas turbine may indicate leakage of oil in the lube oil system either via external piping or internal. It is very dangerous if the cause of high lube oil consumption is not known. Any form of internal leakage in gas turbine shall be paid serious attention because there are many hot components that could be an ignition source for a potential fire occurrence. Thus, the operation teams are highly recommended to keep the lube oil top rate in their logs and inspect the gas turbine condition inside the enclosure. It is recommended to digitalize this information for trending and analysis purposes.

Boroscope Inspection

The boroscope inspection is mainly applicable to gas turbines. It is a critical inspection to be carried out during the annual maintenance. Gas turbine boroscope inspection is an internal inspection performed by a trained personnel to assess the condition of the gas turbine from the air inlet through to the exhaust, using an instrument specifically designed to examine the gas path via the access ports positioned along the engine. In general, the followings are the benefits of boroscope inspection:

- Inspection of internal passages without major disassembly,
- Alert on degraded conditions so that early intervention can be planned prior to catastrophic failure, and
- Contribution to reduce overall maintenance costs.

In order to perform boroscope inspection effectively, the unit must be washed prior to inspection because one cannot see damage (i.e., sign of crack, coating erosion) with dirty blades. These inspections are one of the critical diagnostic methods for maintaining the equipment. Both rigid and flexible fiberscopes are used in conjunction with specially formed guide tubes to inspect the internal stationary, rotating components and to detect early signs of failure.

The boroscope inspection could provide more insight into incipient failures, which could not be detected via other condition monitoring technique. For instance, minor foreign object damage to the blades, crack initiation and propagation, corrosion, as well as thermal deterioration, mild cracking, or distortion mostly could only be detected via boroscope inspection and will not be apparent in vibration reading or trending. This is especially for aero-derivative engine whereby the vibration measurements are made at the casing, and therefore, minor or moderate blade damages will not be strong enough to be transmitted to casing for detection.

Ultrasonic Inspection

The ultrasonic inspection uses ultrasound translators or convertors which provide the facility to hear high frequency noise from equipment via headphone, gauge the intensity of the sound from a handheld unit and perform analytical spectrum analysis of the sound through the use of software. It is widely used in mechanical rotating equipment applications such as:

- Rolling element bearings—there are condition the bearing may slip inside the housing, producing internal sound.
- Airflow disturbance for fans
- Reciprocating compressors valve leak
- Reciprocating engine suction or discharge valve leak
- Control valve leakages

Thermography

Thermography is the measurement of temperatures using an infrared device. The most common application is in electrical transformers, switchgears, or switchboards, but there are also useful applications for rotating equipment. It detects surface temperature difference and hot spots. The most common areas that are applied are:

1. Reciprocating valves. High-temperature indicates valve passing or failure.
2. Cooling fan belts slipping condition may create heat that can be detected by thermography.
3. Bearing temperature problems can be detected based on heat generated.
4. Steam leaks at steam turbine can be determined due to the heat.
5. Gas turbine external surface temperature profile for any form of leaks such as from the fuel piping, burner, or exhaust joints.

Motor Current Signature Analysis

Motor current signature analysis acquires motor current and voltage signals without interrupting production and analyzing the derived signal to identify various faults. Data can be acquired from the motor control panel, enabling easy testing of remote, inaccessible, or hazardous area motors. Similar to vibration analysis, an FFT analyzer is required for converting the signals from time to frequency domain to analyze the spectrum. Some of the capabilities in terms of failure detections are:

  i. Stator winding health
 ii. Rotor health
iii. Air gap static and dynamic eccentricity

Apart from electrical condition, it also can be used to determine the mechanical condition such as the followings:

  i. Misalignment/unbalance
 ii. Load issues
iii. System load and efficiency
 iv. Bearing condition

Therefore, it is very suitable where there is no possibility to apply other condition monitoring techniques such as vibration, lube oil analysis, and thermograph for motor-driven units. Examples of critical application for rotating equipment are submersible pumps.

Performance Monitoring

This technique often related to efficiency of the equipment or comparison against the baseline performance of the equipment. In rotating equipment, it often applied to centrifugal pumps, centrifugal compressors, and gas turbines. However, with advanced technology, it has also been applied to reciprocating compressors using pressure–volume (*P-V*) diagram.

As for centrifugal pumps, operating parameters such as suction pressure, discharge pressure are used to calculate the pump head and with the measured flow, it is plotted against the manufacturer tested curve to determine performance degradation to the equipment. This is one of the indicators to determine the need for major maintenance or overhaul. The NFPA 25 has stipulated the need for annual fire water pump performance testing. Performance degradation of more than 5% shall need further investigation to determine the cause of degradation. The typical centrifugal pump performance curve is shown in Fig. 4.

Centrifugal compressor performance is also determined based on a similar method. The gas turbine performance is a bit more complex as it is given by the manufacturer based on the heat rate versus ambient air temperature graph. The heat rate (commonly in kJ/kW h) is the reciprocal of gas turbine efficiency and therefore is calculated based on:

$$\text{Heat Rate} = \frac{\text{Energy Consumption of Gas Turbine}}{\text{Output Work}} \quad (1)$$

where the energy input or consumption (in kJ) is based on fuel consumption, and the output work is in the unit of kW h.

In some cases, the OEM provides the tested graph with relationship between specific fuel consumption against the exhaust gas horsepower during the factory performance testing. In order to execute the performance evaluation at site, one needs the torque meter and the fuel metering device to plot the parameters into the performance graph for comparison. The other methods to monitor the gas turbine performance are as follows:

- Plot the turbine outlet exhaust gas pressure (corrected) versus exhaust gas temperature (corrected). Plot this graph after the overhaul as a baseline, compare the operating parameters against the baseline reference, and note for any changes. This graph is good indication for overall gas turbine performance monitoring. An example of the graph is shown in Fig. 5.

- Plot the compression ratio of the axial air compressor of gas turbine versus speed (corrected) after the overhaul or commissioning. Use this as the baseline for future operating data, and note for any significant changes. This is a good indication on the performance of the axial air compressor and can be used as an indicator for blade washing activity.

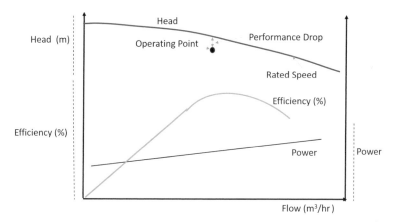

Fig. 4 Typical centrifugal pump performance curve

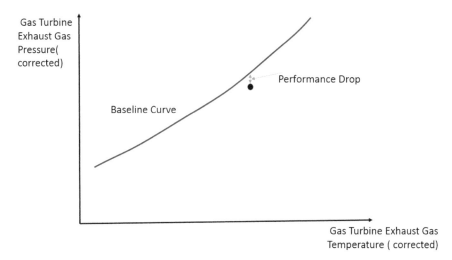

Note: The term corrected value for gas turbine means the value is converted into ISO condition (The ISO standard condition is also specified in API 616 standard). The simple way to convert to the corrected value is by dividing the relevant data with $[(\text{Temperature}_{\text{ambient}} (°C) + 273.15) /288.15]$

Fig. 5 Typical GT exhaust gas pressure (corrected) versus gas turbine exhaust gas temperature (corrected)

- Plot the baseline graph variable inlet guide vane (VIGV) position against the engine speed (corrected). Use this as the baseline for future operating data, and note for any significant changes. This is a good indication of any malfunction in the VIGV system such as electronic and mechanical problems.

Start Reliability and Trip Data

Not every equipment startup will be successful. There are various reasons that a system will not allow for successful start. Therefore, the startup reliability needs to be monitored and tracked to understand the start behavior of the equipment. Startup reliability is tracked by:

$$\text{Start Reliability } (\%) = \frac{\text{No. of Successful Start}}{\text{No. of Start Attempt}} \quad (2)$$

Tracking this information will indicate the reliability of the start system, and later, this information can be used for benchmarking for comparison and system improvements.

In addition, it is imperative to monitor the number of trips that occur on the machine. It is an industry-known fact that there are trips which are caused by instrumentation called spurious trips. This occurs purely due to faults with instrumentation device and no other genuine mechanical abnormal condition. However, every trip causes additional mechanical wear out to the equipment due to transient condition. For instance, typically every trip may add additional 10 h into the life of gas turbine. Thus, the information needs to be tracked and monitored.

Besides this, for equipment such as gas turbines, the operators are recommended to log the time duration taken from start of button till the equipment reaches its normal operating condition. Similarly, this is also applicable when the equipment requires a normal stop. Any deviation in the time will indicate that some internal failures such as bleed valve or variable inlet guide vane (VIGV) problems, gas turbine compressor fouling, changes in gas turbine load, and others which are taking place in the equipment. This is one of the simplest and most powerful data for condition monitoring of the gas turbine.

Implementation of Condition Monitoring Program in Industry

In industry, condition monitoring is implemented as tactical strategy to assess equipment health condition. As equipment becomes more complex and critical in nature, the industry cannot afford low-reliability equipment. Condition monitoring techniques becomes the fundamental technology to transform the old practice on run to

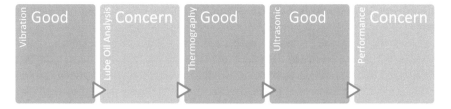

Fig. 6 Condition monitoring health status for a gas turbine

failure to predictive maintenance. Equipment insipient failures are detected much earlier before the equipment needs to be stopped on mechanical failure. The early identification avoids severe damages and costly repair but also helps for a proper planning on tools, manpower, and spares needed to execute the maintenance.

Vibration and lube oil analysis are the most popular techniques used in oil and gas industry for rotating machinery. The other techniques are becoming popular and gradually being applied for better prediction on equipment health and failures. These techniques are performed by specifically trained personnel responsible for their respective condition monitoring skills.

Integrated Approach for Real Time Condition Monitoring

Since condition monitoring activities are performed by a specific trained skill group, the reports on equipment health generated by the individual based on the techniques that have been used. It is important that all the health information that has been obtained using various techniques is being integrated with equipment health, and failure prediction is performed in an integrated manner. This would be the most holistic approach in using the condition monitoring techniques. Figure 6 shows an example of condition monitoring health status.

Risk-Based Maintenance Approach for Rotating Machinery

As for rotating equipment, condition monitoring techniques become a critical input for risk-based maintenance or inspection. Risk-based inspection or maintenance is the maintenance activity driven by risk rather than purely engineering-based data such as condition monitoring. It is basically used to prioritize maintenance by focusing maintenance resources toward equipment that has highest risk, while equipment with lower risk is given less priority. Traditionally, the engineering team is responsible for equipment uptime, and they focus on reducing the probability of failure without considering the consequence of failures such as production loss and maintenance cost.

A risk assessment is performed by combining the probability of failure against consequence of the failure to determine the risk ranking. The risk ranking is then plotted against time, and the equipment with the highest risk is given priority by providing more resources to execute the maintenance and failure prevention activities. By doing so, the risk is mitigated to as low as reasonable and practically possible and the resources are optimized to reduce cost and meet business needs.

The probability of failure is obtained from the followings:

   i. Historical failure data analysis such as previous maintenance data which can be used in Weibull Analysis to determine the probability of failure with time. This would be the appropriate method to determine probability of failure.
  ii. Previous maintenance findings as described below in this section.
 iii. Inspection and testing activities such as boroscope inspection.
 iv. Condition monitoring data on degradation and failure prediction.
  v. Failure analysis on equipment types. This analysis focus on failures such as related to vibration, degraded performance, seal failure, and their probability of failure with respect to time.
 vi. Remaining life analysis as it is important on some critical components such as gas turbine blades which has creep or fatigue life.

It should be noted that Weibull analysis gives a good co-relation on failure probability with respect to time on replaceable component or repairable components that can be restored as good as a new one.

The risk assessment is performed by plotting the probability of failure with consequence based on risk assessment matrix as shown in Fig. 7. The probability of failure for rotating equipment casing can be determined by using technology such as ultrasonic thickness measurement on components of the rotating equipment such as casing and nozzle. Using these techniques, one needs to inspect the casing of rotating equipment at every opportunity maintenance and record the wear pattern and rate such as erosion and corrosion. The casing thickness is measured at reference point to assess the degradation and prediction of failure. This is a critical aspect of rotating machinery maintenance and inspection as, generally, the rotating personnel gives priority for the rotating components such as rotor while the casings are often neglected.

In addition, during preventive maintenance such as overhaul, one needs to record the observation during equipment dismantling. Critical information that needs to be gathered includes the journal bearing clearance, wearing clearance, bushing clearance for centrifugal pump. These readings are again compared against the last as-installed or as-built reading to determine the wear rate. This wear rate information for the equipment can be used as a guide to predict the future condition of the machine based on operating hours. This information shall be used again to determine the next scheduled maintenance at an optimal interval rather than a fixed time-based approach. This information also provides the probability of failure with respect to time and therefore shall be used as an input in risk assessment.

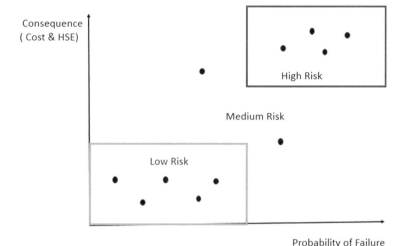

Fig. 7 Determination of risk through the probability of failure versus consequence (Example)

Rotating equipment switching policy is very crucial to provide information on hidden failures. This activity can be considered as testing activity for standby equipment to surface any problems for keeping this equipment on standby for long periods. This information will provide data on probability of the system failure into the risk assessment matrix.

Remote Monitoring and Diagnostics (RM&D)

Remote monitoring and diagnostics (RM&D) is originated by the original equipment manufacturer (OEM) to take some responsibility over the equipment reliability by monitoring and trending the parameters transmitted online from the equipment via customer data infrastructure. A group of experts working remotely at the OEM diagnostic center is employed to analyze the data and predict the equipment health condition. This system is intended to provide early warning sign for any equipment internal failures so that action can be taken at a very early stage. By analyzing the data, the OEM shall provide the action that needs to be taken to the operators. OEMs who have access to big data information from similar equipment around the world could use the data and knowledge for prediction of failure in more accurate and precise manner.

As technology progresses, automated technology is available as software packages in the market. They can be purchased and used even without extending the equipment data to OEM. This technology saves internal man-hours on analysis as the system can automatically read the deviation, early warning signs and notify the operator on the next course of action. The algorithm for the automated monitoring and analysis can be developed in-house if sufficient expertise and knowledge of the equipment are available in an organization.

The prescription function of the technique can be based on a list of common failure modes associated with the equipment. The prescriptive action shall be established to rectify each failure mode. This will help the diagnosis and prescription to be performed in an automated manner. It reduces the man-hour utilization drastically on monitoring and analysis while the workforce can focus on other useful activities such as the development of long-term solutions.

Summary

The elaborated condition monitoring techniques are keys for predictive maintenance approach and understanding of the risk of equipment during operation. The advancement of technology has made condition monitoring data digitized thus enabling integrated analysis to be performed for a high degree of prediction. Condition monitoring technologies made equipment to be inspected based on the health diagnosis, and with risk-based maintenance, the maintenance is prioritized to provide reliability and business value to the organization. As time progresses, there could be other advanced technologies which will be introduced to expand the diagnosis of equipment during operation.

Bibliography

1. B. Quesnel,*Best Practice for Using Oil Analysis in Lubrication Management* (Noria Publication, Machinery Lubrication, 4/2017, 2017)
2. A.J. Smalley, D.A. Mauney, Risk based maintenance of turbomachinery, in *Proceedings of the 26th Turbomachinery Symposium*, Texas A&M University, Turbomachinery Laboratories (1997), pp. 177–187
3. B.M. Basaraba, J.A. Archer,*IPT's Rotating Equipment Training Manual—Machinery Reliability & Condition Monitoring* (IPT Publishing and Training, 1995)
4. API Standard 616,*Gas Turbine for Petroleum, Chemical and Gas Industry Services*, 5th edn. (Jan 2011)
5. API Standard 610,*ISO 13709: 2009, (Identical) Centrifugal Pumps for Petroleum, Petrochemical and Natural Gas Industries*, 7th edn. (Sept 2010)

Turnaround Activities of Rotating Machinery for Mechanical Engineers

M. B. Baharom and M. Azrul Nizam B. M. Zahir

Rotating machinery turnaround is very crucial to manufacturing plants to ensure that equipment is being serviced prior to being used in a long continuous service operation. This chapter discusses on the preparation, execution, and post-turnaround activities of rotating machineries. The contents of knowledge in this chapter are suitable for turnaround planners, mechanical engineers, and contractors. Examples of good practices in turnaround activities and samples of work execution methods are presented in this chapter.

Introduction

Some industrial plants must be continuously running in order to ensure that their productions meet the targets and demands. If the plants are forced to shut down due to whatever reasons, there are production losses and the expenditure of restarting-up the plants will be very costly. Therefore, in order to ensure that the plants are continuously running, backup systems (e.g., rotating machinery) which normally come in pairs are introduced. In this case, one of the units will be in operation while the other one will be on standby. While being on idle, maintenance activities can be carried out on the standby units, hence ensuring that the plants can be continuously operated.

Unfortunately, implementing backups for every equipment or systems is very costly. The analogy is similar to having a pair of duplicated industrial plants for production. Such practice is also not advisable because some equipment cannot

M. B. Baharom (✉)
Universiti Teknologi PETRONAS, Seri Iskandar Perak, Malaysia
e-mail: masrib@utp.edu.my

M. Azrul Nizam B. M. Zahir
PETRONAS Chemicals MTBE, Pahang, Malaysia

© The Author(s), under exclusive licence to Springer Nature Singapore Pte Ltd. 2019   23
S. A. Sulaiman (ed.), *Rotating Machineries*, SpringerBriefs in Applied Sciences
and Technology, https://doi.org/10.1007/978-981-13-2357-7_2

be kept idle for a very long time. In addition, it is not economical to shut down an operating equipment for maintenance activities if its service period has not yet expired. Also, frequent switching of equipment for services may cause plant upset. Therefore, industrial plants need to have 'turnarounds' to ensure effective equipment service renewal.

Turnarounds are scheduled events where industrial plants are completely shut down for a certain period of time so that maintenance activities, modification, and repair works can be carried out on relevant equipment. The importance of having turnarounds is to ensure that the rotating machineries are in healthy conditions, and hence, plants can be operated safely. In this chapter, it is assumed that rotating machineries are the heart of the plants and the turnarounds are mainly intended for them. The required activities for the execution of rotating machinery turnarounds are the preparation, scheduling, execution of tasks, monitoring activities, and post-turnaround activities. The discussions in this chapter are mainly intended for mechanical engineering-related activities.

Classifications of Rotating Machinery

In general, rotating machinery (RM) can be described as mechanical systems or components that possess kinetic energy which are used to move other objects or materials. They can be classified into three main criteria such as the driver, driven, and accessories. The brief descriptions about RM criteria are described in the following subsections.

Drivers

The drivers for RM can be powered by steam, fuel, process liquids, electricity, etc. The main purposes of the drivers are to drive other RM or serve specific purposes such as generating electricity. A few examples of RM drivers which responsibilities belong to mechanical engineering personnel are as follows:

  i.   Steam turbines: powered by steam
 ii.   Gas generator (GG): powered by fuel gas
iii.   Expander: powered by process fluids
 iv.   Internal combustion engines: powered by diesel, petrol, etc.

Fig. 1 Photograph of a motor-driven multi-stage pump

Driven

Driven RM is powered by their drivers as previously mentioned. A few examples of driven RM which responsibilities belong to mechanical engineering personnel are as follows:

  i. Pumps: driven by motors, steam turbines, process fluids, etc.
 ii. Compressors: driven by steam turbines, GG, process fluids, etc.
iii. Fans: driven by motors, ICE, process fluids, etc.
 iv. Blowers: driven by motors.

Figure 1 shows a photograph of a motor-driven multi-stage pump.

Accessories

Rotating machinery accessories are the equipment or systems required for effective operations of either the drivers or the driven equipment. Shown in Fig. 2 is a photograph of a motor-driven blower where power is transmitted through a belt. A few examples of RM accessories are:

Fig. 2 Photograph of a motor-driven blower where power is transmitted through a belt

  i. Gearboxes: for gear reduction or step-up from the driver to the driven
  ii. Oil accumulators: for the lubrication systems of both the driver and driven
     equipment for pressurized oil storage that activate when there is a drop in oil
     pressure
  iii. Dampener: to stabilize discharged flow of positive displacement pumps or com-
     pressors
  iv. Couplings: to connect driver output shafts to the driven input shafts.

Preparation Activities for Turnaround of Rotating Machinery

The cost of conducting turnarounds is very expensive, and hence, mismanagements
are definitely not desirable. The direct cost of turnaround includes labors, consumables, spare parts. The cost is so significance that it should be included in the plants
yearly maintenance budget. In order to ensure that the turnarounds are smoothly

executed without any additional incurred costs, the following subsections explain the tasks that are required to be identified and prepared.

Types of Rotating Machinery

The turnaround activities to be prioritized are the activities that cannot be performed when RM are in normal operation. Routine activities such as preventive maintenance while RM is in operation can also be executed, but it should not be in the critical path. The types of turnaround activities to be executed are maintenance work that requires equipment shutdown, repair work or parts replacement, and modification work either on the RM or other systems related to the equipment.

The maintenance work that requires equipment shutdown normally involves expensive/critical equipment that are without standby units. An example of it is the replacement of air filter of a gas generator. Some RM parts such as compressor or pump mechanical seals should be replaced after a certain running hour as recommended by manufacturers. In some cases, minor repair work may be required for rotating machinery, for example, hairline cracks on casing. During the turnaround, modification works such as modifying equipment piping configurations can also be carried out. Therefore, opportunities should be taken to perform all of these activities during turnaround. The turnaround planner should gather all the information from relevant parties and plan them for the turnaround activities.

Site Visits

Turnaround planners should conduct site visits to identify locations and access the conditions of RM involved in the turnaround. Among the criteria to look for are the requirements for scaffolding, insulation removal, and lifting (i.e., crane). Although the information about them can also be found from the piping and instrumentation diagram (P&ID) and equipment drawings, it is highly recommended to conduct the site visits as the actual conditions of RM need to be confirmed. For example, some modifications might have been carried in the surrounding area or on the equipment by other parties, but they were not updated in the drawings.

Equipment Drawings and P&ID

The P&ID and equipment drawings should be made available for each RM to be serviced or overhauled during the turnaround. If the turnaround is the first ever to be conducted, the planners should produce a file for each equipment. If such files are already available from past turnarounds, the planners can make use of them, but they must make sure that such drawings are the most updated versions as some modifications might have been carried out during the past turnarounds.

Job Method Instructions for Rotating Machinery Overhaul

Prior to preparing the job method instructions (JMIs), it is best to classify RM based on their criteria and types. This procedure is very important because the JMIs for the RM belonging to certain specific criteria and type may be the same or similar. It should be noted that although certain equipment may belong to different classifications, they may also have the same JMI. Examples of RM classifications are shown below:

a. Steam turbines:

- Single stage
- Multistage

Note: They may be classified further based on types of flows, blades, pressures, etc.

b. Pumps:

- Centrifugal
- Positive displacement

Note: They may be classified further as overhung, submersible, inline, etc.

An example of a JMI for an overhaul of a turboexpander is shown in Table 1. In general, JMIs can be prepared based on RM manufacturer's recommendation which is usually available in RM operation manual.

In the JMIs, the activities to be carried out should be arranged in sequence. Removal process involves dismantling of parts, while reinstallation process is the reverse of the removal process. The first part to be removed will be the last part to be reinstalled. If the descriptions in the RM manual are not very clear and the jobs to be carried out are the first time, JMIs can be prepared based on equipment assembly drawings and P&IDs.

Generally, the work activities for JMIs can be classified into two criteria, namely pre-shutdown and post-shutdown activities of RMs. The pre-shutdown activities are mostly related to preparation work. Examples of pre-shutdown work activities are as follows:

i. Mobilization of tools, lifting tools, spare parts, etc.
ii. Isolation of process media, lubrication systems, and anything else related to the equipment.

The post-shutdown activities are the core work to be conducted on the equipment. Examples of activities should include the followings:

i. Step-by-step methods to install and reinstall parts/components in sequential order

Table 1 An example of a JMI for an overhaul of a turboexpander

No	Work activities	Duration (h)	Manpower				Man-hour
			Supv	Tech	Other division	General worker	
Pre-shutdown work							
1	Liquid and hydrocarbon freeing activity	2					
2	Isolate turboexpander	1					
3	Isolate lube oil supply to gearbox	1					
4	Remove instrumentation attached to gas turboexpander	1	1	1			2
Shutdown work							
5	Isolate seal gas (N_2) supply to seal ring (by Opn and Mech)	1					
6	Drain out all (used) lube oil from lube oil tank into drum	3		1			3
7	Remove coupling cover	2	1	1			2
8	Remove tubing connector on seal ring (Inst)	2			2		4
9	Perform pre-alignment check (record all reading for final report)	3	1	1		3	15
10	Check equipment levelness (record all reading for final report)	3	1	1		3	15
11	Remove igv actuator (igv travel gap must be locked)	2	1	1		3	10
12	Remove expander suction and discharge piping (including insulation)	5	1	1		3	25

(continued)

Table 1 (continued)

No	Work activities	Duration (h)	Manpower				Man-hour
			Supv	Tech	Other division	General worker	
13	Remove expander case diffuser	4	1	1		3	20
14	Unbolt and remove front section of gas turboexpander	5	1	1		3	25
15	Remove expander wheel and mech. seal.	4	1	1		3	20

    ii.   Special instructions on specific or general tools to be used
   iii.   Special instruction on the removal of special parts
   iv.   Special instruction for the coordination of removal and reinstallation of parts/components belonging to other divisions
    v.   The number of required people to perform the tasks in hierarchy order
   vi.   The estimated duration for each task
  vii.   The number of man-hours required for each task including the manpower from other division. In most common practices, one day is equivalent to 8 h
 viii.   Instruction on what to be done on the new/old parts or components
   ix.   Special instructions for checking of tolerances and performing alignment.

    Figure 3 shows a photograph of lifting of a large compressor for maintenance, which fall under the category of post-shutdown activities.

Job Inspection Checklists

The job inspection checklists (JICs) for RM service or overhaul are prepared based on the required activities to be performed on the RM. The general guidelines for the preparation of JICs are given below:

    i.   To identify main tasks to be performed on an equipment
   ii.   To list all the sub-activities for each main task which need to be done before, while and after performing work on the equipment
  iii.   To come up with a method for each sub-activity
  iv.   To assign the owner or a responsible person for each sub-activity
   v.   To assign a quality controller (QC) or an inspector for each sub-activity.

    An example of JMI and the checklist of an overhauling work of a steam turbine is illustrated in Table 2. Figure 4 shows an inspection of gas turbine blades.

Table 2 An example of a JMI and the checklist of an overhauling work of a steam turbine

No	Activities	Method (work/measurement/inspection)	Approval		Remarks
			QC	Owner	
Removal and reinstallation of top casing					
Before top casing removal					
1	Pre-alignment check	Dial gauge			
2	Measure DBHE/DBSE	Inside micrometer			
3	Mechanical bump test	Dial gauge			
4	Check and record hub travel	Measure/depth caliper			
5	Inspect governing valve chamber	OEM checklist			
6	Check radial bearing clearance	OEM checklist			
After top casing removal					
7	Check rotor to inner parts clearance	OEM checklist			
Removal and reinstallation of rotor					
After removal of rotor					
8	Secure vibration probe sensor and bearing seat area	PVC tape			
9	Clean turbine rotor and internal	Special cleaning agent			
10	Turning gear condition	Photo			
While/after rotor installation					
11	Bearing clearance check	Measure/feeler gauge			

(continued)

Table 2 (continued)

No	Activities	Method (work/measurement/inspection)	Approval		Remarks
			QC	Owner	
12	Record rotor clearance	OEM checklist			
13	Mechanical bump test	Dial gauge			
14	Measure DBHE/DBSE	Inside micrometer			
15	Final alignment	Dial gauge/laser			

Job Safety Analysis

Job safety analysis (JSA) can be prepared based on potential hazards from the nature of tasks to be performed, environmental conditions of workplace, materials related equipment processes, etc. An example of JSA for an overhaul of a centrifugal pump

Fig. 3 Lifting and maintenance activities on a large compressor

Fig. 4 Inspection on gas turbine blades

is shown in Table 3. The main intention of JSA is to identify potential hazards and recommend the safest way to execute tasks. In addition, when JSA is prepared, it should follow safety standards such as OSHA. In the JSA table, the persons/groups responsible to take actions should be mentioned as to ensure that the persons are aware of their obligations. It is strongly recommended that JSA be prepared through brainstorming sessions consisting of maintenance, technical services, and operation personnel. The list of activities to be performed can be obtained from JMIs. For each of work activity, the team will identify the hazards and then propose the best method to execute the task. For example, one of the identified hazards from the previous example of JMI, namely 'Remove expander wheel and mechanical seal' is lifting hazards. In this case, all personnel are required to follow the safety procedures for lifting.

Bill of Materials

Prior to performing turnaround activities on any RM, the turnaround planners must ensure that the required bills of materials (BoM) are available for the affected equipment. BoM consists of spare parts and consumables that require replacement during equipment servicing. If such parts or consumables are not available or out of stock, they should be ordered immediately and the delivery period should be before the start

Table 3 An example of a JSA for an overhauling work of a centrifugal pump

Job steps	Possible hazards	Effects	Precautions/preventive measures/corrective actions	Actions by
Preparation activities 1. Mobilize equipment to site	(1.1) Poor understanding of working area or location	(1.1) Accident/incident during mob.to site	(1.1.1) Discuss with supervisor to confirm work location	Rotating
2. Manual unloading of Tools (grinder and welding set) and materials	(2.1) Incorrect lifting	(2.1) Pinching	(2.1.1) Lift in a correct manner with proper PPE	Rotating
	(2.2) Material falls	(2.2) Personal or foot injury	(2.2.1) Wear proper safety shoe	Rotating/workshop
			(2.2.2) Use correct lifting tools	Rotating
Maintenance execution 3. Apply PTW for works	(3.1) Mis-communication	(3.1) Accident/incident	(3.1.1) Supervisor to liaise with operations and only to allow work to proceed when permitted by operations	Rotating/utility
	(3.2) Safety violation	(3.2) Accident/incident	(3.2.1) To comply to PTW requirement	Rotating work leader
4. Grinding	(4.1) Fire due to spark and LEL level not safe	(4.1) Accident/fire incident	(4.1.1) Ensure and request for LEL check at every 4 h with permanent LEL meter installed at all time	Rotating/utility/safety

(continued)

Table 3 (continued)

Job steps	Possible hazards	Effects	Precautions/preventive measures/corrective actions	Actions by
			(4.1.2) To use fire blanket and standby fire extinguisher at working area	Rotating/workshop
5. Removal of pump top casing from skid. Lifting of rotor from lower casing	(5) Falling material	(5.1) Equipment damage	(5.1.1) Ensure use of proper lifting tools and ensure correct crane rating used (refer load chart)	Rotating/chain block
		(5.2) Personnel injury	(5.2.2) barricade and put signage at lifting area to ensure nobody enters the lifting area	Rotating
6. Installation of Pump rotor to lower casing. Installing pump top casing	(6.1) Falling material	(6.1.1) Equipment damage	(6.1.1) Ensure use of proper lifting tools and ensure correct crane rating used (refer load chart)	Rotating/chain block
		(6.1.2) Personnel injury	(6.1.2) Barricade and put signage at lifting area to ensure nobody enters the lifting area	Rotating
	(6.2) Insufficient lighting	(6.2) Accident/incident	(6.2) Ensure sufficient lighting prepared before lifting	Rotating

of turnaround. The list of items for BoMs can be obtained from equipment manual. The BoMs should consist of notes whether the spare parts/consumables are custom made by only specific manufactures or commonly available from any manufacturers as long as the specifications are the same. The executors of turnaround activities should ensure that all items stated in the BoMs are available prior to performing job tasks.

Preparing Blind Lists and Size of Blinds

The process as well as lubrication inlets and outlets of RM are connected to external piping through flanges. In order to isolate RMs from piping that has been depressurized and purged, blinds are required to be installed in between equipment and rotating flanges. Ideally, the thickness of blinds should follow standard specifications stated in ASME or API standards. However, spreading of flanges and inserting blinds in between them are not easy tasks. Therefore, the thickness of blinds may not follow the standard but it should be able to withstand the piping purged pressure. If a pressure test is required to be performed on RM casing, the thickness of blinds should follow test specifications.

The preparation of blind lists and size of blinds are important because the information can be used for planning of turnaround activities. The tasks of installing blinds may not belong to RM personnel, but they are responsible for such activities because the equipment belongs to them. Hence, RM personnel should coordinate with other divisions for the tasks to be carried out.

Identifying Assisted Access

This task includes aspects such as scaffolding erection/dismantling, lifting cranes, insulation removal, and reinstatement. Some RM may be big in size or they may be located at elevated locations. Hence, scaffolding structures are required for access to perform work activities. RM personnel may liaise with other divisions for the erection and dismantling of required scaffoldings. Due to the big sizes or being installed at elevated locations, lifting work is required for removal and reinstallation of parts. Therefore, the arrangement of lifting cranes is required. Depending on process conditions, RM may be insulated and RM personnel may liaise with other divisions for the removal and reinstatement of such insulation materials. It is the responsibilities of RM personnel to identify all assisted accesses as parts of preparation for turnaround works.

Table 4 Examples of stakeholders and information for scheduling purposes

No	Stakeholder	Information
1	Equipment owners (e.g., engineers or supervisors)	JMI, JSA, BoM, and manpower availability
2	Equipment operators (e.g., operation or process engineers)	Duration for RM shutdown and starting up
3	Materials owners	Availability of parts, delivery period and conditions of parts

Turnaround Work Schedule

Effective turnaround work schedule is very crucial in successfully implementing RM turnaround activities because it clearly specifies intended tasks to be carried out during a specified time frame. The preparation of work schedule should be done by turnaround planners and schedulers who coordinate all the tasks to be done from all involved parties consisting of plant stakeholders such as RM owners, plant operators, materials owners (warehouse) and the schedulers themselves. The information to be included and timing of the schedules can be obtained from stakeholders, and a few examples are mentioned in Table 4.

A responsible person or owner of the schedule should be appointed and in most common practices the person with the highest role, e.g., plant manager. As the turnaround period should be as short as possible, the duration is set based on critical path activities. The critical path is an activity or a sequence of activities which takes the longest time to complete in the turnaround schedule. It is also the shortest possible time taken to complete the selected task. For the case of rotating machinery, the critical path for RM turnaround is usually the longest period taken to service an equipment plus the time taken to shut down and start up the plant. For any RM servicing which does not belong to the milestones activities, the tasks can be performed in parallel with other activities. When preparing the turnaround schedule, work precedence should be identified so that activities can be executed smoothly.

Preparing Contract Specification

It is important for a company to weigh their capabilities in terms of manpower, resources, and financial as well as insurance matters in order to decide whether the RM turnaround activities are to be executed by their own manpower or contracted out. The company also needs to decide whether to take full responsibility on tasks or to hand over some responsibility to contractors. Such decision determines the statements that should be written in the contract specification. For example, if a company is willing to take its own responsibility, it may only hire general workers or specialists to assist their technical personnel to perform tasks. In this case, work instructions and decision making come from the company, and any after-service faults are borne by the company. If a company wants to hand over some responsibility to the contractors, it can specify in the contract specification that the contractor needs to supply suitable and sufficient manpower to perform tasks based on existing JMIs or contactor's newly proposed JMIs and to complete work within the specified schedule. The previous decision is also influenced by the availability of resources such as special tools, cleaning equipment and inspection instruments belonging to the company. For expensive and complicated equipment that can be considered as the heart of the plant, it recommended that the company award the contract to equipment manufacturer's certified service providers and the after-service should be covered under insurance. In this case, the contact specification specifies the contractor to perform certain tasks, produce its own JMIs and complete work within specified schedule.

The contract specifications should be detailed and accurate so that contractors can provide reasonable price quotations. If the contract specifications are too general or unclear, contractors may impose hidden charges. During the actual work execution, some unexpected activities may be encountered. In this case, it is recommended that the company request contractors to provide the cost of manpower per day, the lump-sum price for performing certain tasks, equipment rental per day, etc. The activities to be conducted by contractors prior and post-equipment servicing should also be spelled out in the contract specification.

Work Execution

The turnaround should only be conducted when all the preparation work has been completed, and confirmation has been made regarding the availability of spare parts and other resources. Ideally, during the turnaround, all work should be executed based on what has been planned within the specified planned schedule. The most important of all is to ensure that all safety rules are being adhered so that working places, workers, and equipment are in safe conditions. As unexpected phenomenon may happen during the turnaround, all activities should be monitored to ensure that activities are being executed smoothly within the specified period.

Safety

All the safety rules and regulations should be followed by all personnel at all levels who are involved in turnaround activities. Turnaround safety officers should conduct a walkabout around workplaces to ensure that job executioners are obeying safety rules. Superiors should ensure that all personnel under their responsibilities follow safety regulations. All personnel should wear personnel protective equipment such

Fig. 5 Lifting of a gas turbine generator. Note that the workers are wearing proper PPE while performing work

as safety helmet, steel-toed and oil-resistant shoes, and hand gloves during servicing of RM (Fig. 5).

Prior to any work execution, a work permit issued by the RM custodian should be made available and hung at workplace. The permit should clearly mention the types of work to be carried out, i.e., either 'hot' or 'cold' work. For hot work permit, activities that produce fire spark are permitted while for cold work, no work-producing spark is allowed and only non-spark tools are permitted. For the case of RM with large casing, vessel entry work permit is required.

Among the most important safety rules to be followed for RM servicing work is the 'lockout' and 'tag out' procedures. RMs are powered by some form of energy sources; therefore, they must be isolated and lockout before any work can be carried out on them. Examples of equipment energy resources include electrical, steam, pneumatic, and thermal. The term 'lockout' simply means that a lock device is placed on energy-isolating device so that the equipment cannot be operated. The term 'tag out' means that a display is showed at the tag-out device warning anyone not to operate the equipment until the tag-out device is removed. The tasks of performing 'lockout' and 'tag out' are normally performed by other divisions and the responsible persons who are in charge of servicing the RM should ensure that such tasks are conducted prior to commencing any work.

Monitoring Activities

Monitoring of turnaround activities are required to be conducted so that appropriate actions can be taken to ensure all the activities are on track. A few recommended actions to be taken are as follows:

(a)   Measurement of Key Performance Index (KPI)

The measurement of KPI can be performed by comparing the actual work against the planned activities. Among the criteria for measurement include costing, time taken, and quality of service.

(b)   Monitoring of Work Progress

Work progresses need to be tracked by comparing the actual executed work against the planned work schedules. A software can be used to monitor the activities, and an alert will be triggered if there is a delay in the schedule. Constant frequent meetings are required for updates on work status. Depending on the criticality of turnaround period, meetings can be held as frequent as one in the morning and one in the evening or once a day in the morning. It is recommended that the schedule is owned by the highest authority in the turnaround organization. Due to unexpected phenomenon, the planned schedule may be altered and such decision should be made by the highest authority. The decision on the change of schedule should be based on priority and business requirements.

(c)   Monitoring of Work Quality

The quality of work should be monitored to ensure that all activities follow certain quality standards. This is to ensure that the plant is in healthy conditions during future operation. Although inspection activities are included in the inspection checklists, monitoring of such activities by certain parties is required.

(d)   Monitoring of Resources and Unexpected Happenings

Despite all the planning work, the actual activities may differ due to some unexpected happenings. For example, when an equipment is opened up for an overhaul work, the internal parts may be found to be defective possibly due to corrosion. Hence, repair work may be required and this may cause a delay in the schedule. There may also be problems with availability of resources and manpower. For examples, some equipment may be malfunction due unexpected phenomenon or some workers may suddenly fall sick due to food poisoning.

Post-Turnaround Activities

The activities to be carried out after the turnaround are crucial because there will be another turnaround to be conducted in the future. It should be expected that the execution of turnaround activities should be improved as the planners and executioners have gained some experiences. The following activities are required to be performed for the closeout of past turnaround:

 i. To record and file the actual number of manpower, man-hours and cost to perform certain activities
 ii. To record the actual steps taken to perform certain tasks
 iii. To gather and file equipment history such as damage, service done, and causes of damage
 iv. To be indulged in continuous learning activities from past experiences
 v. To performed SWOT analysis on the past turnaround activities and performance
 vi. To tackle issues which emerge during the past turnarounds.

Summary

Turnarounds are scheduled events where industrial plants are completely shut down for a certain period of time so that maintenance activities can be carried out on relevant equipment to ensure that they are in healthy conditions for the plants to be safely operated. For ease of planning activities, RM can be classified into drivers, driven, and accessories. The preparation activities for RM turnaround are as follows:

a. To identify prioritized activities to be conducted
b. To conduct site visits
c. To ensure availability of equipment drawings and P&ID
d. To prepare:

     i. Job method instructions (JMI)
     ii. Job inspection checklists (JIC)
     iii. Job safety analysis (JSA)
     iv. Bills of materials (BoM)
     v. Blind lists and size of blinds

e. To identify assisted access

Effective turnaround work schedule is very crucial to ensure that RM turnaround activities can be effectively executed and all the works can be completed within stipulated time. The inputs from all stakeholders must be obtained in the preparation of the schedule. In preparing contract specification, a company must weigh their capabilities in terms of manpower, resources, and financial as well as insurance matters in order to decide whether the RM turnaround activities are to be executed by their own manpower or contracted out. The contract specifications should be detailed and accurate so that contractors can provide reasonable price quotations. The turnaround should only be conducted when all the preparation work has been completed and confirmation has been made regarding the availability of spare parts and other resources. All the safety rules and regulations should be followed by all personnel at all levels who are involved in the turnaround activities.

During the turnaround execution, the criteria to be monitored are

a. The measurement of key performance index (KPI)
b. The work progress
c. The work quality
d. The availability of resources and unexpected happenings

Post-turnaround activities must be carried out because there will be another turnaround to be conducted in the future and the execution of the next turnaround must be an improved version from the current one because the planners and executioners have already gained some experiences.

Bibliography

1. T. Sahoo, *Process Plants—Shutdown and Turnaround Management* (CRC Press, Taylor and Francis Group, Boca Raton, 2014)
2. T. Lenahan, *Turnaround, Shutdown and Outage Management—Effective Planning and Step-by-Step Execution of Planned Maintenance Operations*, 1st edn. (Butterworth-Heinemann, Elsevier, 2006)
3. H.P. Bloch, F.K. Geitner, *Practical Machinery Management for Process Plants—Machinery Component Maintenance and Repair*, 3rd edn. (Gulf Professional Publishing, Elsevier, 2005)
4. C.D. Reese, *Occupational Health and Safety Management—A Practical Approach* (CRC Press, Taylor and Francis Group, Boca Raton, 2008)
5. D.C. Richardson, *Plant Equipment and Maintenance Engineering Handbook* (McGraw-Hill Education, 2014)

Localization of Gas Turbine Maintenance by Malaysian Gas Turbine Owners and Operators

Mohd Amin Abd Majid and Ismady Ismail

Gas turbines are important equipment for power industry. Local gas turbines are all imported and of various capacities. Among important issues that need to be addressed is to localize the maintenance expertise. Initiatives are being undertaken by the local turbine owners and operators to reduce the imports of the components through localization of refurbishments and parts replacements. This study is to investigate the current practices of localization of gas turbine maintenance works by the Malaysian industry. The study covers gas turbine maintenance practices of the local gas turbine owners and operators and the initiatives undertaken to localize gas turbine maintenance expertise. Potential areas which could enhance the localization of gas turbine maintenance expertise are also highlighted.

Introduction

Gas turbines (GTs) are among major equipment used for generating power. They are of various capacities. They are easy to install and to start. In Malaysia, the GTs are installed and operated by Tenaga Nasional Bhd. (TNB), Independent Power Producers, and PETRONAS Operating Units (PETRONAS OPU). They are being operated either for open cycle, cogeneration cycle, or combined cycles [1, 2]. The GTs that are installed in the Malaysian industry were imported from various countries. Although the GTs have been installed and operated for some times in Malaysia and can be categorized as among common equipment, the maintenance works, inspection, refurbishments, and parts replacement are very much dependent on the original equipment

M. Amin Abd Majid (✉)
Universiti Teknologi PETRONAS, Perak, Malaysia
e-mail: mamin_amajid@utp.edu.my

I. Ismail
TNB Repair and Maintenance, Selangor, Malaysia

S. A. Sulaiman (ed.), *Rotating Machineries*, SpringerBriefs in Applied Sciences and Technology, https://doi.org/10.1007/978-981-13-2357-7_3

43

manufacturer (OEM). Thus, maintenance costs are among the main costs of owning and operating the GT. If the maintenance, parts refurbishment, and replacements can be localized, then the operating costs can be reduced.

This study investigated the potential of localization, the maintenance expertise, and parts manufacturing of the main GT components. The approach adopted covers three main aspects. First is survey of the current maintenance practices, practiced by the industry. Second is the identification of the main components that are normally refurbished and replaced. Third is the identification of the maintenance works and components manufacturing that are feasible for localization.

GTs are widely used in industrial sectors of Malaysia. Among the main users are TNB and PETRONAS OPU. The installed capacities vary. Most of the GT installed by TNB are of utility category, while PETRONAS OPU installed the mini category [1, 2]. The utility category varies from 12,500 to 265,000 kVA, while the mini category is with capacities ranging from 650,00 to 10,000 kVA [3]. All the three categories of turbines use natural gas, either diesel or fuel oil as fuel. The service outages for micro, mini, utility are 2 years, 8 months, and one-and-half years, respectively. The operating RPM among the three categories of turbines differs substantially. Microturbines operate at 70,000 RPM, and miniturbines operate at 15,000 RPM. The utility turbines operate at very much lower RPM, which is 1800 RPM. The efficiency of utility turbines is 37%; it is lower at 32% for microturbines and 30% for miniturbines. The purchase and installation periods for the three categories of turbines differ greatly from each other. The microturbines require only about a week to purchase and install. Miniturbines require about 2 months to purchase and install. The utility turbines require substantially long period about one to 2 years to purchase and install.

It a known fact that all the GTs installed in Malaysia are imported from various countries among which are USA, UK, France, Germany, and Japan. The maintenance works and parts replacement of the GT are highly dependent to OEM or OEM local representatives. Due to this scenario, the GT spares are mainly imported.

Among the main factors that affect maintenance planning are facilities and capabilities and economic issues [4]. Four factors under facilities and capabilities which are linked to maintenance management are knowledge and experience, recommended maintenance program, data collection and analysis, on-site maintenance capabilities and availability of replacement parts [4]. On the economic issues, three factors that are highlighted are cost of downtime, repaired reliability, and life cycle cost. All these factors need to be considered in localization of GT maintenance expertise and works.

In this study, the focus is to investigate the extent of local involvement in the GT maintenance works and parts refurbishment and replacement of the GT. Specific focus is on the practices adopted both by TNB and PETRONAS OPU since they are the main owners and operators of the GT.

Gas Turbine Maintenance and Inspection

In order to ensure the GT can operate with optimum availability and reliability, appropriate periodic inspection, repair, and replacement of parts, in particular the hot gas path parts, schedule is required [5]. The hot gas path parts include the combustion liners, transition pieces, turbine nozzles, and turbine buckets. Other parts that require maintenance services are control devices, fuel metering equipment, and other station auxiliaries. In terms of inspection and repair requirements, Knorr et al. [5] suggested to include establishing cyclic inspection patterns covering from minor works to major overhauls. The inspection patterns either operational or teardown should be optimized to reduce unit outages and maintenance costs while maintaining maximum availability and reliability. The operational inspections are used as indicators of the general condition of the equipment and as guides for planning the maintenance program. These inspections can be regarded as standby, running, combustion, hot gas path, and major.

Knorr et al. also highlighted on standby maintenance for GT used in intermittent service. For this situation, starting reliability is essential. This category of maintenance includes routine servicing of the battery system, changing of filter, checking oil and water levels, cleaning relays, and checking device calibrations. The services can be performed during off-peak periods in order to avoid interrupting the availability of the GT. It is also important to perform periodic test run inspection.

Other point raised by Knorr et al. is related to running inspections. This is continuous in nature, and the inspections consist of the general and continued observations made while a unit is in service. As a guide, it is recommended that an attended station is probably observed every shift, or at least daily. The unattended, continuous duty machine will probably be observed on a 1- to 4-week basis, depending upon its accessibility. The intermittent duty unit will be observed every 5–10 starts or at a minimum once a month. Operating data is to be recorded in order to enable performance and maintenance assessments of the equipment.

Navrotsky [6] reported an initiative on maintenance plan (MP) for GT SGT-600 with a target to minimize downtime to increase availability. The target planned outage is to extend the maintenance intervals from 20,000 to 30,000 h. The MP was designed to reduce downtime of the inspection and site activities in terms of three areas, namely extension of shift work, reduction of Level-A inspection from 3 days to 1 day, and improvement of the maintenance processes and tools. The MP was initially implemented on GT SGT-600 installations with base load operation profile and latest component design. The extension of the maintenance interval from 20,000 h had enabled the operator to save two overhauls, performing three intervals instead of five. At the same time, the duration of the remaining inspection was reduced. In total, the whole life cycle and the availability increased by about 1%.

HITACHI had published guideline on maintenance inspection schedule for HITACHI GT [7]. The schedule categorizes inspection into three categories, namely combustion inspection, hot gas path inspection, and major inspection. For combustion inspection based on 2-year operation, recommended inspection interval is 16,000 h for turbines using natural gas fuel and 12,000 h for turbines using oil fuel. For hot gas path inspection based on 4-year operation, recommended inspection interval for turbines using natural gas is 32,000 and 24,000 h for turbines using oil fuel. For major inspection based on 8-year operation, the recommendation is 64,000 h of inspection interval for turbines using natural gas. If the turbines used oil fuel, the inspection interval is 48,000 h.

Degradation of GT

Kurz et al. [8] identified four mechanisms that caused degradation. The four degradation mechanisms are fouling, hot corrosion, corrosion, and abrasion. Kurz et al. [8] also highlighted the causes of the degradation and the following recommendations to mitigate the degradation mechanisms:

a. Fouling caused by adherence of particles to airfoils and annulus surface. The adherence in terms caused by oil and water mist, which resulted in a build-up of materials that caused increased surface roughness and to some degree changed the shape of the airfoil. Recommended mitigation is to do regular offline and online washing.
b. Hot corrosion caused by loss of deterioration of material from flow path components, which was due to chemical reactions between the components and certain contaminants. This causes formation of scales to the components. Recommended mitigation is protection through an oxide scale.
c. Corrosion caused both by air contaminants and by fuel combustion-derived contaminants. It could be controlled with filtration—with proper selection of filter media.
d. Abrasion caused when a rotating surface rubbed on stationary surface. The recommended mitigation is cleaning or washing of the engine. Adjustment, repair, or replacement of components are also recommended [9].

GT Maintenance and Refurbishment Practices Locally

GT owners and operators are aware on the importance of implementing proper maintenance practices and parts refurbishments and replacements in order to ensure optimum GT performance. The current approach adopted by the local GT owners and operators to achieve this objective is to relay the service of OEM and OEM local representatives which are licensed by the OEM. The following three of the five cases highlighted this scenario.

Case A: Maintenance Activities at a Cogeneration Plant of Malacca Refinery Company (MRC)

The plant is a cogeneration plant. Five units GE GT of 25 MW (ISO standard) are installed at the plant. The power generated by the GT is used for refinery processes and the plant requirements. The plant adopted the following maintenance schedules.

The first combustion inspection is 12,000 h. This involves checking and replacement of combustor, fuel nozzle, and liner. The second maintenance schedule is at 24,000 h. This involves inspection and parts replacement for the combustor and inspection of turbine side. The third maintenance schedule is at 36,000 h. This is a shutdown for combustor inspection similar to the 12,000 h, schedule shutdown. The next maintenance is at 48,000 h. While at 96,000 h, the contract for maintenance works is renewed. All the scheduled maintenance works are contracted to OEM. The contract includes manpower and refurbishment as well as replacement of original spare parts.

GE has collaboration with Sapura for transfer technology in terms of GT maintenance covering local manpower and refurbishments. The scope for refurbishment covers turbine blade, nozzle, bucket, combustion liner, and transition piece. The works are done at the workshop in Shah Alam and at GE Keppel, Singapore.

RAMACO has approached MRC for GT maintenance. Since the contract with GE is ongoing and will last another 5 years, the GT maintenance could not be packaged to RAMACO.

Case B: GT Maintenance Works by RAMACO

The main activities of TNB REMACO are:

- Operations and maintenance (O&M) of power plants such as TNB Prai (Siemens H class), combined cycle power plant, and Manjung 5 Coal Fired Power Plant. Including O&M contracts in Kuwait and Pakistan.
- Power plant maintenance field services.
- Refurbishment of power plant equipment and machinery such as IGT hot gas path parts.
- Test and diagnostic for electrical and mechanical (vibrations and NDT).
- Projects such as power transmission lines.

Among GT, maintenance activities undertaken are:

- Extension of maintenance and inspection intervals for IGT hot gas path parts. The inspection intervals are:

  - Combustion inspection is carried out at 24,000 fired factored hours (FFH) instead of 8000 (FFH) previously.
  - Life extension of stationary and rotating IGT parts from OEM recommend scrapping of parts at 72,000–96,000 FFH.

- These practices enable TNB REMACO and TNB to make substantial saving in terms of maintenance inspection costs and capital parts purchases.
- In terms of localization of GT parts, current initiative is to manufacture the GT exhaust system through reverse engineering. The project is being undertaken in

Table 1 Heat treatment results for Inconel 617 by OEMs and ISPs

OEM and ISP	A	B	C	D	E
Component	HGC, IL	MC	IC	IC	IC
Heating rate	70–300 °C/h	80 °C/h	<300 °C/h	Unknown	Unknown
Soaking temperature	$1125 \pm X$ °C	$1125 \pm X$ °C	$1120 \pm X$ °C	1176 °C	1170 °C
Soaking time	4–5 h	1 h	>3 h	1 h	Unknown
Cooling rate	70–300 °C/h to 650 °C/h, then unrestricted	Air cool	<300 °C/h to room temperature	Furnace cool	Unknown

collaboration with a partner from Holland. A local fabrication shop in Ipoh, Perak, is used for fabrication works.

- Another ongoing initiative is to investigate the reverse engineering of pumps used in power plants. The main constraint faced is on the special materials required for the pumps. Due to stringent operating conditions, special materials such as super duplex stainless steel are required for pumps in corrosive environments. However, there are very few companies in Malaysia which can cast the special materials that can meet the level of quality specified for oil and gas API standards.
- In addition, initiative is being undertaken to develop heat procedure for Inconel 617 material. Inconel 617 is used for the construction of IGT structural combustor component by OEMs. The study is undertaken due to contradicting results received for heat treatment cycle from OEMs and ISPs (Sulzer Turbo and Inpirio) as shown in Table 1.

The project is expected to achieve the following benefits;

- To understand the reason, OEMs/ISPs carry out such heat treatment cycle.
- To avoid using proprietary information from OEMs.
- To prove that TNB REMACO has developed the heat treatment procedure in-house and not merely duplicating others.
- The methodology of the research can be used to develop heat treatment procedure for other materials in the future.
- The HT procedure is to support component refurbishment procedure.

Although there are local manufacturers which manufacture gears and other rotating parts which can be used for GT, the asset owners in particular the main plant owners and GT users are not ready to use the parts. This is due to high risk involved.

Case C: Maintenance Practices for 2 × 4.2 MW Gas Turbines at a Cogeneration Plant

The cogeneration plant generates power and chilled water. The power generated is used for an academic institution and the plant. The chilled water is generated by steam absorption chillers and electric chillers. Preventive maintenance is conducted for the

GT every 4000 running hours interval. The scope of maintenance works covers the following:

i. Mechanical—engine offline water wash, borescope inspection, and filters changeout.
ii. Instrument—calibration on transmitters and switches.
iii. Electrical—electrical motor and MCC inspection and other related minor maintenance.

All the parts change are through OEM except the air intake filters that are locally sourced from CAMFILL.

Case D: Adaptive Neuro-Fuzzy Inference System for Performance Health Monitoring of Offshore GT
This initiative is to investigate the application of adaptive neuro-fuzzy system for monitoring of offshore GT. The basis is that predicting whether or not a GT is inclined to faults provides useful help for determining the required preventive action before failure happening. System identification is a discipline that learns the behavior of the healthy engine and employs it to predict the fault proneness. Adaptive neuro-fuzzy inference system (ANFIS) was developed and compared to the artificial neural networks (ANNs) for the purpose of GT performance identification. Three system identification Bank of Networks (B-Ns), each corresponding to seven variables that are commonly measurable on most twin shaft industrial GT engines, are developed. The accuracy of the trained B-Ns is analyzed using the healthy performance data of an industrial 18.8 MW open-cycle offshore GT. Making a comparison between the gained results from ANFIS and two various ANNs, models revealed that ANFIS model is able to forecast various performance parameters with higher correlation coefficient and smaller MAPE values [10].

Case E: Planned Maintenance for PETRONAS Integrated Complex
PETRONAS Integrated Project (PIC) is in the process of development of a new perspective on the plant maintenance. The initiative of plant maintenance is to adopt digital asset maintenance ICT work stream. This is a new approach of working and most likely an extensive adoption of predictive analysis. This includes the cogeneration plant at the integrated complex. The cogeneration plant is installed with GT with total generating capacity of about 800 MW. Analysis is also on algorithm for predictive analysis.

Additional Feedbacks on GT Maintenance
The following points related to GT maintenance were highlighted:

- There is no local manufacturer for GT critical parts such as compressor or turbine blade, fuel nozzle.
- AIROD in Subang has capacity to manufacture certain parts.
- TNB REMACO has license to repair certain GE frame six parts. The Connaught Bridge Power Plant also does some reverse engineering for their Siemens GT parts.
- Part like roller bearing could be manufactured locally as there is SKF Factory in Nilai.

- Hose and filter for hydraulic or oil system could also be locally manufactured.
- Air intake filters already produced locally.
- SULZER (a third-party service provider) produces many GT or pump parts in Indonesia as they have bigger market.

Discussion

The maintenance practices on GT in Malaysia industry are time based and very much dependent on support provided by OEM. The scope of maintenance covers inspection, parts refurbishments, and replacements. The owners and operators of GT are confident and satisfied with the services provided by the OEM and the OEM local representatives. There are initiatives to increase the localization of GT maintenance in terms of scope of inspection, parts refurbishment, and replacement. This is packaged in collaboration of OEM with local companies to undertake specific maintenance and parts refurbishment.

From feedbacks by GT owners and operators, there was no mention on the importance of either offline or online washing, which could reduce fouling and abrasion. Filtration and coating which could address the corrosion [6] were not being given priority. Hence, local GT owners need to consider incorporating these maintenance scopes in maintenance agenda in order to enhance the life of GT. An initiative to adopt condition based maintenance along with digital maintenance is ongoing. The initiative is commendable and will lead to the improvement of maintenance management GT. This initiative should be adopted as a common maintenance management practice for the local GT owners and operators.

Malaysia manufacturing industry is well established. There are substantial resources in terms of manufacturing facilities and pool of technical manpower available which could be employed to undertake structured GT maintenance. The missing element is structured framework and coordination. Hence, it is important to address these issues which could enhance the localization of GT maintenance management.

Summary

Malaysian industry installed a substantial quantity of GT of various capacities. Most of the capacities are medium and industrial categories. The industry is very much involved in GT maintenance with OEM support. Some not comprehensive and unstructured initiatives are ongoing to localize the GT maintenance. To address this issue, a framework is required which could integrate the initiatives for localization of GT maintenance. The framework should be integrated and encompass the GT owners, operators, OEM, and the local OEM agencies.

Acknowledgements The authors wish to acknowledge the assistance of Mr. Haider Abu Bakar of Malaysian Refining Company, Mr. Shahairul Fahizan and Mr. Khairuddin Ibrahim of Makhostia Sdn Bhd for providing maintenance information.

References

1. S.A. Sulaiman, M. Amin Abd Majid, *Overview of Gas District Cooling in Malaysia, in Gas District Cooling in Malaysia* (Lambert Academic Publishing, Saarbrücken, Germany, 2011), ISBN 978-3-8465-5441-8
2. Wikipedia, List of power stations in Malaysia, https://en.wikipedia.org/wiki/List_of_power_s tations_in_Malaysia. Accessed 18 Sept 2017
3. H. Lee Wills, W.G. Scott, *Distributed Power Generation, Planning and Evaluation* (Marcel Dekker, Inc., New York, 2000)
4. R. Hoeft, E. Gebhardt, *Heavy Duty Gas Turbine Operating and Maintenance Considerations* (General Electric Company, Gas Turbine Division, 1993)
5. R.H. Knorr, G. Jarvis, Maintenance of industrial gas turbines, in *ASME 1975 International Gas Turbine Conference and Products Show*, Paper No. 75-GT-93, pp. V01BT02A031, 8 pages, 1975
6. V. Navrotsky, Gas turbine performance and maintenance continuous improvement, in *VGB Conference, Gas Turbines and Operation of Gas Turbines 2013*, Siemens Industrial Turbomachinery AB, Sweden, 11–12 June 2013
7. Hitachi Ltd., *HITACHI H-25 & H-80 Gas Turbine* (Bucharest, 2013)
8. R. Kurz, K. Brun, Gas turbine tutorial-maintenance and operating practices effects on degradation and life, in *Proceedings of the 36th Turbo Machinery Symposium*, Texas A&M University, Turbomachinery Laboratories, 2007
9. I.S. Diskinchak, *Performance Degradation in Industrial Gas Turbines*, ASME Paper 91-GT-228, 1991
10. M. Tahan-Bouria, M. Muhammad, Z.A. Abdul Karim, Adaptive neuro-fuzzy inference system for performance health monitoring of industrial gas turbines, in *Proceedings of the 2016 International Conference on Industrial Engineering and Operations Management*, Kuala Lumpur, Malaysia, 8–10 Mar 2016

Probabilistic Life Cycle Costing: A Case Study of Centrifugal Pumps

Ainul Akmar Mokhtar, Freselam Mulubran
and Masdi Muhammad

Life cycle costing (LCC) has been gaining attention in industries as a decision-making tool for the management of assets. LCC is generally recognized as a valuable tool in making an optimal decision considering the total cost of ownership rather than just the initial acquisition cost. The deterministic LCC model is commonly used in many plants; however, the deterministic model only inherently encompasses uncertainty factors, i.e., the economic issue alone, and cannot practically and effectively handle the ambiguous other uncertainties such as changes in interest rate, cost of production loss per hour due to unexpected failure, and labor cost per failure, to name a few. The life cycle cost of a repairable system is closely linked to and highly influenced by its maintenance cost which includes its reliability, maintainability, and maintenance support. Thus, to incorporate these uncertainties into LCC, one needs to consider the application of the reliability engineering principles to evaluate the probabilistic nature of the equipment failures and repairs.

Overview of Life Cycle Cost

LCC is one of the quantitative tools which is increasingly being used in the industry to make various types of decisions in managing assets by estimating the costs that will be incurred over the whole life of the asset or the total cost of ownership. The estimates of cost elements used in LCC analysis are typically prone to uncertainty, and deterministic models can lead to inaccuracy and thus can affect the decision made. Therefore, to minimize the inaccuracy of the analysis, it is important to model the relevant uncertainties of the input data. However, there is still a large gap in the industry practices. A research conducted in Finland by Korpi and Ala-Risku [1]

A. A. Mokhtar (✉) · F. Mulubran · M. Muhammad
Universiti Teknologi PETRONAS, Perak, Malaysia
e-mail: ainulakmar_mokhtar@utp.edu.my

© The Author(s), under exclusive licence to Springer Nature Singapore Pte Ltd. 2019   53
S. A. Sulaiman (ed.), *Rotating Machineries*, SpringerBriefs in Applied Sciences
and Technology, https://doi.org/10.1007/978-981-13-2357-7_4

found that 83% of manufacturing industries under study used the deterministic LCC analysis and only 17% used the probabilistic model. It was an indication that the analysis assumes the deterministic behavior of the input parameters.

Dhillon [5] stated that some important reasons why there was an increasing trend in the use of LCC in industry are due to the high variation between the product procurement cost and cost of ownership. This was reflected in the rising cost of maintenance and operation costs. According to Stewart et al. [3], several studies on cost inferred that cost of ownership of engineering system varies from 10 to 100 times the original cost of acquisition of the system.

In a process of asset management, it is important to properly estimate the present value of acquiring a system and the cost of ownership. This will help in the review and update of the system's LCC financial element for both procurement and operation life time. In the operation phase of equipment, the system often runs under rigorous conditions, thus leading to wears, aging, erosion, etc. These rigorous operating conditions cause the equipment/system to fail unexpectedly, which results in large financial loss, accidents, deaths and environmental pollution [3]. The definition for LCC varies [4] by different scholars; for example, Waghmode et. al. [8] defined LCC as "The costs incurred on a product during its whole life cycle" and Okafor [9] defined it as "The summation of the total costs of a product, process, or activity discounted over its lifetime".

LCC can be conducted in all life cycle of equipment or a system, either in the design phase of a new equipment/system or in the operation phase of existing equipment/system. The main reasons for doing LCC in the operation phase are that:

- In design phase, the LCC is typically conducted based on theoretical concept and based on some assumptions.
- During the production phase, the life cycle cost estimates made on the equipment will change because of either equipment obsolescence, change in design or integration with other equipment/systems.
- Analyze differences between forecasting and actual costs.
- Assist to identify the potential areas of cost saving and cost drivers.
- Implement management control.

Life Cycle Costing Model

The general life cycle cost models are not tied to any specific system or equipment. Dhillon [5] proposed different general LCC models, and three of them are shown below:

Model 1: In this case, the equipment or system life cycle cost is divided into two main parts: recurring cost and nonrecurring cost. Thus, the system or equipment life cycle cost is expressed by:

$$LCC = RC + NRC \tag{1}$$

where RC is the recurring cost and NRC is the nonrecurring cost. The recurring costs include operating cost, inventory cost, support cost, manpower cost, and maintenance cost, and the nonrecurring cost includes costs associated with procurement, installation, qualification approval, research and development, training, reliability and maintainability improvement, and support.

Model 2: In this case, the equipment or system life cycle cost is divided into three main parts: procurement cost, initial logistic cost, and recurring cost. Thus, the system or equipment life cycle cost is expressed by:

$$LCC = C_1 + C_2 + C_3 \tag{2}$$

where C_1 is the acquisition or procurement cost, C_2 is the initial logistic cost, and C_3 is the recurring cost. The initial logistic cost is composed of one-time costs such as the cost of procurement of new support equipment, not accounted for in LCC solicitation and training, the cost of existing support equipment modifications, and the cost of initial technical data management. The three main components of the recurring cost are operating, management, and maintenance cost.

Model 3: In this case, the life cycle cost is expressed as:

$$LCC = C_1 + C_2 + C_3 + C_4 \tag{3}$$

where C_1 is the research and development cost, C_2 is the production and construction cost, C_3 is the operation and support cost, and C_4 is the retirement and disposal cost. Estimating the total LCC requires breakdown of the asset into its constituent cost elements over time. The level to which it is broken down will depend on the purpose and scope of the LCC study. In this case study, Model 3 is adapted and classified into six main categories: acquisition, installation, operation, maintenance, production downtime, and decommissioning. The general LCC is shown by:

$$LCC = C_{aq} + C_i + C_{op} + M_c + P_L + D_c \tag{4}$$

where LCC is the life cycle costing, C_{aq} is the acquisition cost, C_i is the installation cost, C_{op}, is the operating cost, M_c is the maintenance cost, P_L is the production loss due to downtime, and D_c is the decommissioning or disposal cost.

Acquisition Cost

The general expression for the acquisition cost is given by:

$$C_{aq} = \sum_{j=1}^{j=n} C_j \qquad (5)$$

where j is the components of the initial acquisition costs, for example the initial purchasing price, bidding process, inspection, transportation, initial spare parts inventory, training, auxiliary equipment, product planning, engineering design, product test and evaluation, software's used, design documentation, training, raw materials, and all the research development and manufacturing phases.

Installation Cost

The pump installation and commissioning costs include the foundations, grouting, connecting of process piping, connecting electrical wiring, connecting auxiliary systems, equipment alignment, flushing of piping, and performance evaluation at start-up. The care and effectiveness in executing the installation activities will have a great impact on the subsequent reliability during the life cycle of the pump, hence affecting the maintenance and downtime costs. However, for this case study of pump, installation cost is assumed to be included in the acquisition cost. The installation cost of all the activities is given by:

$$C_i = \sum_{i=1}^{i=m} \left[(t_a \cdot n_p \cdot C_{li}) + C_{ti} \right] \qquad (6)$$

where C_i is the cost of installation, t_a is the estimated time for each activity, n_p is the number of labor needed for each activity, C_{li} is the labor cost for installation, and C_{ti} is the cost of tooling for installation.

Operation Cost

The operating costs are classified into two categories which are the pump energy consumption and the labor costs related to the operation of a pumping system. Pump energy consumption, which is part of the operation cost, is often one of the larger cost elements and may dominate the total life cycle costs, especially if pumps run more than 2000 h per year. Operating costs are also labor costs related to the operation of a pumping system. These vary widely depending on the complexity and duty of the system. Regular observation of how a pumping system is functioning can alert operators to potential losses in system performance. Performance indicators include changes in vibration, shock pulse signature, temperature, noise, power consumption, flow rates, and pressure.

The high-impact cost drivers in the operation phase are the number of operation hours, personnel, and cost of energy. By integrating these factors, the operation cost can be estimated by:

$$C_{op} = t \cdot [(C_e \cdot kW) + C_l] \tag{7}$$

where C_e is cost of energy ($/kWh) and C_l is cost of labor per hour.

Energy consumption is calculated by gathering data on the pattern of the system output. The cost of energy for pump can be estimated by:

$$C_e = C_{pw} \left(\frac{Q \times H \times \text{s.g.}}{366 \times \eta_p \times \eta_m} \right) \tag{8}$$

where C_{pw} is cost per input power (in $/kW), Q is the pump flow rate (in m^3/h), H is the pump head (in m), η_p is the pump efficiency, η_m is the motor efficiency, and s.g. is the specific gravity.

Maintenance Cost

One of the main factors which affect the reliability of the system is proper maintenance. Uncertainties arise from maintenance cost determination, because the failure of the system can happen stochastically. The general equation of maintenance cost is given by:

$$M_C = C_C N + C_P N_P \tag{9}$$

where M_C is the maintenance cost, C_C is the cost of repair, C_P is the preventive maintenance cost, N is the number of failures, and N_P is the number of preventive maintenance.

Number of preventive maintenance, N_P, can be estimated based on the type of maintenance policy adopted by the plant. There are two types of preventive maintenance: scheduled and condition-based. The cost of scheduled or time-based maintenance is fixed; however, the cost of condition-based maintenance is probabilistic. In order to determine N_P, the condition of the equipment or machine will be determined based on the trend of its failure.

The corrective maintenance is conducted whenever there is a failure. The cost of repair, C_C, is estimated by the activities performed:

$$C_C = C_{sp} + C_t + (\text{MTTR} \times l \times n) \tag{10}$$

where C_{sp} is the cost of spare part for repairing a failure and C_t is the cost of tools. If the pump is repaired without replacing any parts, C_{sp} is going to be zero. MTTR is the mean time to repair, l is the cost per labor, and n is the number of labor for each activity. Number of failure, N, for repairable item can be expressed by:

$$N = \frac{T}{\text{MTBF}} \tag{11}$$

where T is the total life span of the system and MTBF is the mean time between failure.[1]

Production Downtime Cost

The cost of failure of the system associates not only with maintenance cost but also with downtime or loss of production cost. Due to downtime, the system is unavailable, which results in the loss of production and can be calculated by:

$$P_{\text{L}} = D_t Q C \tag{12}$$

where P_{L} is the loss of production, D_t is the cumulative downtime due to failure, Q is the production per hour, and C is the cost of production per unit.

Decommissioning Cost

The disposal cost can be estimated from all activities conducted. The major activities identified include pump disassembly, separation, material recovery, and dumping. The cost drivers associated with these activities such as time to disassemble the pump, weight of the pump, distance of transportation, quantity of material to be recovered, quantity of material to be dumped, and numbers of persons involved have been identified. The pump disposal cost can be estimated by:

$$D_c = \left[C_h \sum_{i=1}^{i=n} T_l \right] + [(C_{tr} \cdot x \cdot H_r) - (C_{scr} \cdot H_r)] + \left[\left(C_{kdp} \cdot y \cdot H_{du} \right) + (C_{du} \cdot H_{du}) \right] \tag{13}$$

where D_c is the pump disposal cost, C_h is the hourly cost of disassembly, T_l is the disassembly time for component, C_{tr} is the cost of transportation of the recovered

---

[1] This equation is applicable only for constant rate of occurrence of failure.

material, x is the distance over which the recovered material is to be transported, H_r is the weight of the recovered material, C_{scr} is the scrap rate, C_{kdp} is the cost of transportation of the decomposed parts, y is the distance over which the decomposed parts are to be transported, H_{du} is the weight of the decomposed parts material, and C_{du} is the cost of decomposed parts. In this case study, the decommissioning cost is not included.

Probabilistic Life Cycle Costing Through RAM Analysis

Probabilistic life cycle costing requires defining the probability distribution for every uncertain variable such as time to failure and downtime due to repair and mainte-nance. It is necessary to consider uncertainties because of the uncertain operating characteristics. In probabilistic life cycle costing, the number of failures for estimat-ing the maintenance cost and the cumulative downtime for downtime cost can be estimated using the reliability, availability, and maintainability (RAM) concepts. In this case study, the other cost components, acquisition cost, installation cost, and decommissioning cost are assumed to be the same as the deterministic life cycle costing.

Reliability and maintainability are essentially analytical in nature and character-ized a probabilistic process. When equipment or system is on its operation phase, it will fail and be repaired to restore to its operating condition. The number of failures, when it fails, and the time to repair are determined through reliability and main-tainability analysis in which the maintenance data, failure data, and repair time are considered. Through reliability analysis, how many times the equipment will fail in the operational life span can be estimated and subsequently the cost of failures can be derived.

Similarly, the rate of recovery of a system to its normal working condition will also affect the costs incurred. There is equipment which incurred multiple failures over its life span and returned back to the service after repairs such as pump. MTBF and MTTR form the critical cost parameters for reliability of repairable system.

In a study by Hwang [11], they analyzed the initial system configuration to achieve certain performance parameters based on RAM and life cycle cost by developing a static model. They also developed a time and failure truncated model for system RAM test. The acquisition, support, and maintenance costs were taken into account while applying it in a production facility using the developed life cycle cost model.

Overview of the Case Study

A case study of the amine pump systems in the gas-processing plant is used to illustrate the concept of probabilistic LCC. The amine pump systems are the critical parts of the gas-processing plant which is considered as the core of the business. The

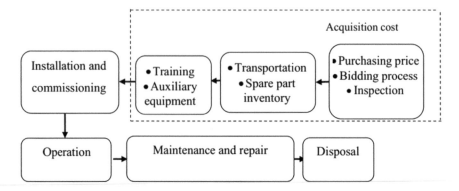

Fig. 1 Pump life cycle stages

primary function of gas-processing plant is to separate natural gas into sales gas, ethane, propane, butane, and stabilized condensate.

The gas-processing plant has two amine pump systems, where each system has five units of amine circulation pumps to the main absorbers. Four out of five pumps (Pump A, B, C, D) are driven by steam turbines and hydraulic power recovery turbines (HPRT). The fifth pump, Pump E, is driven by an electric motor and is used as a spare.

Pump life cycle refers to sequential phases from the need identification to the disposal. The major life cycle stages of a typical pump are shown in Fig. 1.

For this case study, the LCC analysis mainly focuses on the acquisition cost, maintenance cost, loss of production cost due to downtime, and operation cost. The cost estimation for the existing system includes the operation cost, the maintenance cost, and the production loss due to downtime.

Estimation of Acquisition Cost

It is quite challenging to get the same acquisition cost from different vendors. In this case study, a thorough investigation of different vendors was conducted in order to estimate the acquisition cost.

Table 1 shows the operating and cost data for existing and the new pumps proposed to be purchased. This data assists to determine the quantitative requirement for selection of a new pump. All costs are in Malaysian Ringgit. The initial purchase of the pumps includes transportation, bidding process, and installation cost.

From the seven vendors, four of them suggested a price of RM1.18 million for the stated specification in Table 1, while two of them suggested RM1.60 million and the remaining one suggested RM1.20 million. Based on the expert judgment (i.e., based on questionnaires), the probability that the cost of acquisition was RM1.18 million

Table 1 Pump data

Specifications	Existing pump	New pump
Initial cost/pump (RM)	Not available	1,300,000
Pump head (m)	648	750
Rate of flow (m³/h)	1096.92	1136.50
Pump efficiency (%)	83	84
Motor efficiency (%)	81	89
Power consumed (input power)	2771	2750
Energy cost/year	• RM 0.317/kWh for peak hour • RM 0.175/kWh for nonpeak hour	With pump energy consumption 2750 kW/h (data given by the plant)
Design life of pump (year)	20	20

is 0.58, RM1.60 million is 0.28, and RM1.20 million is 0.14. Thus, the expected acquisition cost is:

$$C_{Acq} = (1.18)(0.58) + (1.60)(0.28) + (1.20)(0.14) = RM1.30\,\text{million}$$

Estimation of Operation Cost

The operation cost was estimated using Eq. (7) where the energy cost for pump is a subset of the operating cost. Energy consumption is often one of the larger cost elements. Energy consumption is calculated by gathering data on the pattern of the system output. The pumps are running 24 h a day and 365 days a year which was 8760 h. According to the electric utility company, the high voltage peak/off-peak industrial tariff charges are RM0.317/kWh and RM0.175/kWh, respectively. Cost of energy for pump can be estimated using Eq. (8). The power usage by the existing pump during peak and off-peak time is estimated as:

$$\text{Power usage} = \frac{1096\,\frac{m^3}{h} \times 648.1\,\text{m} \times 0.94}{366 \times 0.83 \times 0.81} = 2713.54\,\frac{kW}{h}$$

Table 2 shows the pattern of the power usage for a day and the probability of running with that pattern, and this is based on the historical data and expert opinion. For example, the probability of running 50% during peak hour and 50% during nonpeak hour (i.e., Scenario 1) is 0.75.

For example, the annual energy cost when running during the day, 50% during peak hour and 50% during nonpeak hour, is (Table 3)

Table 2 Pattern of the power usage for a day and its probability of running

Scenario	Peak hour (%)	Non-peak hour (%)	Probability of running the scenario
1	50	50	0.75
2	60	40	0.20
3	70	30	0.05

Table 3 Annual energy cost for different scenarios

Scenario	Energy cost (RM/year)		
	Peak hour	Nonpeak hour	Total
1	3,767,641.75	2,079,928.41	5,847,570.16
2	4,521,170.10	1,663,942.73	6,185,112.83
3	5,274,698.45	1,247,957.05	6,522,655.49

Annual energy cost (peak hour)

$$= 2713.54 \frac{kW}{h} \times 0.317 \frac{RM}{kWh} \times 0.5 \times 24 \frac{h}{day} \times 365 \frac{day}{year}$$

$$= RM\,3,767,641.75$$

Annual energy cost (non-peak hour)

$$= 2713.54 \frac{kW}{h} \times 0.175 \frac{RM}{kWh} \times 0.5 \times 24 \frac{h}{day} \times 365 \frac{day}{year}$$

$$= RM\,2,079,928.41$$

$$Annual\ energy\ cost = RM3,767,641.75 + RM2,079,928.41$$
$$= RM5,847,570.16$$

Using Eq. (7), the expected total energy cost is therefore RM 5,948,832.96:

Total energy cost

$$= 5,847,570.16 \times 0.75 + 6,185,112.83 \times 0.20 + 6,522,655.49 \times 0.05$$
$$= RM5,948,832.96 \approx RM5.95\ millions$$

Estimation of Maintenance and Production Downtime Costs

A number of improvements have been made for the sustainability of amine pump operation. Among the significant improvement are: minimum flow line installation,

Table 4 Options to increase the pump systems' availability

Option No.	Description
Option 1	To continue with the existing pump system through repair. The LCC of the existing pump is estimated by predicting the availability and reliability of the pump in the future
Option 2	To modify the existing system by changing one or two of the old pumps, which have lower performance, by the new pump. The pump with the highest downtime will be changed with a new pump first and will estimate the improvement that it makes to the system. Similarly, the modification will be conducted for the pump which causes the second highest downtime and so on

pump impeller upgrade, mechanical seal upgrade, and third bearing installation. However, despite all the improvements being made, the availability of amine pump was still a concern. Recent statistics showed that the amine pump system was still struggling to achieve high availability and reliability due to repetitive failure cause by vibration and mechanical seal issues. According to the industry report and the expert opinion, there are two options to consider in order to increase the reliability and availability of the amine pump as shown in Table 4.

In this section, all the two options were investigated using life cycle cost analysis. The LCC was done for the existing system and also for the new pump. The existing system was in service phase (i.e., operating, maintenance, and support). As it is stated in, "Methods and Models for Life Cycle Costing," a report conducted by NATO, the main objective of doing LCC in service phase is to forecast future cost, manage the existing budget, and undertake options analysis where necessary.

Option 1—Continue with Existing Pump System Through Repair

The LCC of the existing pump is estimated by predicting the availability and reliability of the pump in the future. In probabilistic LCC, the number of failures and availability was determined through RAM analysis. To acquire failure data, what constitutes a failure must be defined. Collecting and using data from computer maintenance management system (CMMS) must be carefully considered and pre-planned by thoughtful analysis to prevent accumulating nonuseable data. All data in CMMS had accuracy problems particularly in the age to failure. In order to solve, this CMMS data was integrated with trip record and pump flow rate. Below are the assumptions taken for the extraction of time-to-failure data:

- The past 3-year data was collected and analyzed.
- The failure time was considered in days.
- The flow rate was taken as an average value per day.
- If the flow rate was less than zero, it is assumed as a standby pump.
- The pump is considered running if the flow rate was greater than 600 and considered failed if the failure rate was in between 0 and 600.

- If the failure time recorded was not found in the categories of PPM, SSM, PRM, and REM in the CMMS data, then it is considered as REM.
- The start of failure time was taken as zero from the time in which the pumps completed overhaul.
- If the pump was on standby, it was considered operating.

Based on the OEM's recommendation (operation and maintenance department of the plant), three types of overhaul are typically conducted for the amine pump which are:

- Small inspection (SI) which is performed every year and includes inspection, control, lubrication, cooling, corrosion.
- Major inspection which is performed every 3 years and includes SI, mechanical seal checking, balancing, bearing check.
- Major maintenance is conducted every 5 years and is a major overhaul at workshop.

The number of staffs required for repairing unexpected failure is one internal and two external staffs. Labor cost for an internal staff is RM50 per hour. Since all the pumps are assumed to be the same, the cost will also follow the same distribution.

Reliability growth analysis, using RGA version 10 software, was used to estimate the parameters such as β, λ, η, growth rate, and the number of failures as shown in Table 5. The analysis was done based on multiple system concurrent operating times data which incorporated the failure before and after overhaul for Pumps A, C, and D. For Pumps B and E, the overhaul was done at the beginning of the analysis period. Pumps A and D showed a β value which was less than one which was the indication of system performance and has a positive growth rate. The other three pumps, Pumps B, C, and E, had a negative growth rate which showed that the pumps were degrading. The number of failures for the next 1, 2 and 3 years was estimated.

From the analytical analysis conducted, the CROW/AMSAA test showed that the repair process which the pumps follow was NHPP. The number of failures predicted will be affected based on the repair assumptions taken. It is better to predict a maintenance cost which could be a little bit higher than the real cost rather than predicting

Table 5 Reliability and maintainability data of each pump

		Pump A	Pump B	Pump C	Pump D	Pump E
β		0.701	1.322	1.426	0.726	1.231
λ (in day)		0.110	0.001	0.0017	0.141	0.0018
η (in day)		28.710	196.403	115.292	18.169	170.161
Growth rate		0.299	−0.322	−0.426	0.274	−0.231
Number of failures for the next t year	$t = 1$	7	2	8	10	3
	$t = 2$	11	6	20	17	6
	$t = 3$	15	10	26	23	10

a smaller cost estimate where this will put the decision to be made on the safest side. Therefore, NHPP can be used to model the repair assumption, i.e., as bad as an old assumption.

In order to validate the number of failures, the 3-year actual failure data was taken and compared with the predicted number of failure for the next 3 years as shown in Table 6. The predicted value gave a mean forecast error of -1.8, mean absolute deviation of 2.6, and the tracking signal between -1 and 0.76. As long as the tracking signal is between -4 and 4, assume the model worked correctly.

From time to repair and cost data, the repair and cost distribution were estimated using Weibull++ 10 software. The repair distribution for Pump A, Pump B, and Pump E followed the Weibull distribution, while Pump C and Pump D followed the normal distribution, as it is seen in Table 7. The cost data followed the lognormal distribution with log mean of 6.73 and log standard deviation of 1.49.

Using BlockSim 10 software, the 1000 iterations were performed for 20 years for the amine pumping system and it was found that the availability is approximately 96.8%, with Pump D contributing to the highest downtime for the system. Table 8 presents all the RAM parameters and cost data of each pump. Pump D has poor availability performance followed by Pump C, Pump E, Pump A, and Pump B.

Table 6 Actual and predicted number of failures

Pump	2012	2013	2014	Actual number of failure for 3 years	Predicted number of failure for the next 3 years
A	2	4	6	12	15
B	4	2	1	7	10
C	9	7	16	32	36
D	2	10	10	22	23
E	3	2	7	12	10

Table 7 Reliability analysis

Pump	Distribution		Parameter value
A	Weibull	β	1.56
		$\eta(h)$	124.34
B	Weibull	β	0.89
		$\eta(h)$	40.54
C	Normal	μ (h)	96.26
		σ (h)	57.73
D	Normal	μ (h)	377.85
		σ (h)	162.11
E	Weibull	β	3.31
		$\eta(h)$	201.87

Table 8 RAM parameters and cost data of each pump

	Amine pump system	Pump A	Pump B	Pump C	Pump D	Pump E
MTBF (h)	–	264.78	189	232.83	987.1	750.7
Availability (%)	96.8	58.4	74.9	51.27	37.8	52.6
Downtime (h)	5600.52	72876.68	43802.3	85362.5	108844	83004.7
Expected number of failures	11.385	27	44	37	13	20
MTTR (h)	495.6	2699.13	995.5	2307	8372.6	4150.24
Total cost (RM)—cost of tool and spare part	429,308.13	142,679.12	10,816.01	95,687.26	31,102.19	49,023.55
Cost per failure	–	2284.4	2518.54	2586.14	2392.4	2451.17
Number of labor	–	4	4	4	4	4
Cost per each labor (RM)	–	50	50	50	50	50

Even though the β value for Pump A and Pump D was less than one and has a positive growth rate, the reliability was affected by the repair rate where the mean time to repair Pump D was the highest among all.

Using Eq. (9), the maintenance cost was estimated. Corrective maintenance cost for each pump is shown in Table 9. Pump D contributes 27.5% of the total maintenance cost, followed by Pump C 21.7%, Pump E 21%, Pump A 18.6%, and Pump B 11.2%.

Table 10 depicted the LCC of each pump before and after discount if the plant continues through maintenance. The interest rate used was 3.5%, i.e., interest rate by Malaysian National Bank in 2015. t represents the period the analysis is conducted. Whenever the system is unavailable, the plant loses RM200,000 per hour.

The next analysis focuses on the life cycle of the amine pump if the plant considers changing the pumps which incur higher number of failure and low availability. These are Pump B, Pump C, and Pump A. Before proceeding to analyze this assumption, the LCC of the new pump, which is replacing one of the pumps either with the highest number of failure or with low availability, is estimated in the next section.

Table 9 Maintenance activity and for the next 20 years

Activity level 1	Activity level 2	Pump	Maintenance cost (RM in million)
Maintain centrifugal pump	• Access to the failed component • Diagnosis • Repair/replacement • Verification and alignment	Pump A	14.7
		Pump B	8.9
		Pump C	17.2
		Pump D	21.8
		Pump E	16.6

Table 10 Estimated cost of the existing amine pumps for the next 20 years

		Amine pump system before discount (RM in million)	Amine pump system after discount (RM in million)
Annual operation cost		5.95	83.14
Corrective maintenance cost		79.2	39.8
Downtime cost		1120.02	562.92
Preventive maintenance cost	Cost of small inspection (SI) per year	0.05	0.71
	Cost of major inspection (LI) per 3 years	0.25	0.13
	Cost of major maintenance (LM) per 3 years	2	5.3
NPV of the pump system after discount			692.95

Option 2—Modify Existing System

This option involves modification of the existing pump systems by changing one or two of the old pumps, which have lower performance, by the new pump. It is not possible to get the exact failure rate for the new pump, since it does not have any historical data (i.e., because it never fails). Due to this to establish the global failure rates, the use of international standards database like OREDA is required. The degree of quality of data derived from these data sources differs considerably. In best cases, generic databases such as OREDA provide data that also include engineering and functional characteristics (i.e., system boundaries definitions) to complement the estimation of failure rate in the principal failure mode. In other cases, the information

Table 11 RAM and cost data of the new amine pumps

Parameter		Amine pump system
Initial cost/pump (RM in million)		1.3
MTBF (h)		20.7
Availability		99.5%
Downtime (h)		130
Expected number of failure for the next 3 years		3
Acquisition cost (RM in million)		1.3
Annual labor cost (RM in million)		0.026
Annual energy cost (RM in million)		5.93
Annual operation (RM in million)		5.956
Corrective maintenance cost (RM in million) per year		0.081
Downtime cost (RM in million)		3.9
Preventive maintenance	Replacement of mechanical seal (RM in million/3 years)	0.14
	Lube oil top-up (RM in million/year)	0.01
	Seal top-up (RM in million/year)	0.05
LCC of the pump after discount (RM in million)		23.88

could be very restricted, probably confined to an overall failure rate estimate for general classes of equipment.

Centrifugal pump for gas-processing plant data taken from the handbook will reduce the production loss due to downtime and also maintenance cost. Similar to the existing pumps, one-hour failure will incur RM200,000 production loss. The annual labor cost and the annual corrective maintenance cost of the new pump were taken as 2 and 6.23% of the initial investment, respectively, based on the experience of industries. The data given from the manufacturer of the pumps about the preventive maintenance to consider is shown in Table 11 together with other new pump data. The LCC of the new pump for the next 3 years is found to be RM23.88 million. Operation cost and downtime cost contribute to 78 and 15% of the total LCC.

Reliability and Availability Analysis of the Modified System

The availability of each pump in the existing amine pump system was very low, which is the cause of production loss and maintenance cost. If the plant decides to change all the pumps at the same time, the production will be interrupted and the whole system will be down. The better alternative was changing the existing pumps step by step one pump at time while the others being in operation. It was found from

Table 12 RAM data for the modified amine pump system

Parameter	Existing amine pump system	Amine pump system after changing Pump D only	Amine pump system after changing both Pumps D and C
MTBF (h)	15,388.6	549,216.3	867,326.7
Availability	96.8%	99.9	99.9
Downtime (h)	5,600.52	96.4	10.55

Table 13 Maintenance activity and cost for option 2

Activity level 1	Activity level 2	Pump	Maintenance cost after changing Pump D only (RM million)	Maintenance cost after changing Pumps C and D (RM million)
Maintain centrifugal pump	• Access to the failed component • Diagnosis • Repair/replacement • Verification and alignment	Pump A	13.6	16.2
		Pump B	7.0	12.3
		Pump C	16.3	0.1
		Pump D (new)	2.2	0.1
		Pump E	10.7	0.1

RAM analysis that Pumps D, C, and E have the lowest availability, respectively, and Pumps B, C, and A have the highest number of failures. Since these pumps have the highest downtime that leads the system to be unavailable, changing all the pumps at the same time can lead to very high cost. Therefore, we used one scenario which was changing the pump with low availability step by step. Table 12 shows the availability and downtime of the amine pump system after the replacement of Pump D only and both Pumps D and C.

It is obvious that the availability of a system will be improved if a new equipment is being replaced where the availability will be improved by 3.01% after changing Pump D. The MTBF also shows that the reliability of the pump system was improving.

Even if the reliability and availability were improved by modifying the existing pump, the main factor that affects the replacement is the net present value of the total cost of the existing and the modified amine pump system. The maintenance cost after changing Pumps D and C is shown in Table 13.

After changing Pump D, the highest contributor to the maintenance cost is Pump C with 32.8%. After changing both pumps, Pump D and Pump C, the main contributor to the maintenance cost is Pump A with 56.2%. The LCC for all the pumps is shown in Table 14. The total LCC was reduced from RM692.95 million of the existing system

Table 14 LCC of existing and modified amine pumps

		Existing amine pump system (RM million)	Modified system after changing Pump D (RM million)	Modified system after changing Pumps C and D (RM million)
Acquisition cost		–	1.3	2.6
Annual operation cost		83.14	83.14	83.14
Corrective maintenance cost		39.8	24.98	14.47
Downtime cost		562.92	9.69	1.06
Preventive maintenance cost	Cost of small inspection (SI) per year	0.71	0.71	0.71
	Cost of major inspection (LI) per 3 years	1.06	1.06	1.06
	Cost of major maintenance (LM) per 3 years	5.3	3.18	2.12
Estimated cost after discount		692.95	124.06	105.17

to RM105.17 million with the changing Pump D and Pump C. This shows that the plant rather than continuing with the existing system would be much profitable if it changes two of its pumps.

Summary

The LCC of repairable equipment depends on its reliability, maintainability, and availability. Defining probability distribution requires adequate historical data, which is usually unavailable. Nonhomogenous poison process (NHPP) was used in probabilistic LCC, which includes analysis to produce estimation of RAM measures. In availability analysis, Monte Carlo simulation technique was used to estimate the system availability. The results of RAM measures estimation were then input into the probabilistic LCC.

References

1. E. Korpi, T. Ala-Risku, Life cycle costing: a review of published case studies. Manag. Audit. J. **23**, 240–261 (2008)
2. R. Harris, Introduction to Decision Making, VirtualSalt, 2012. http://www.virtualsalt.com/cre book5.htm

3. R.D. Stewart, R.M. Wyskida, J.D. Johannes, *Cost Estimator's Reference Manual*, vol. 15 (Wiley, New York, 1995)
4. M. Moazzami, R. Hemmati, F. Haghighatdar Fesharaki, S. Rafiee Rad, Reliability evaluation for different power plant busbar layouts by using sequential Monte Carlo simulation. Int. J. Electr. Power Energy Syst. **53**, 987–993 (2016)
5. B. Dhillon, *Life Cycle Costing: Techniques, Models and Applications* (Routledge, 2013)
6. R. Peurifoy, C. Schexnayder, *Construction Planning, Equipment, and Methods* (McGraw-Hill, New York, 2002)
7. O.P. Okafor, Application of Analytic Hierarchy Process (AHP) in the selection of an effective refinery for life cycle cost analysis. Int. J. Eng. Res. Technol. (IJERT) **2** (2013)
8. L. Waghmode, A. Sahasrabudhe, P. Kulkarni, Life cycle cost modeling of pumps using an activity based costing methodology. J. Mech. Des. Trans. ASME **132** (2010)
9. O.P. Okafor, Development of a life cycle cost estimating framework for oil refineries. Thesis submitted in partial fulfilment of the requirements for the degree of Master of Science by Research M.Sc. thesis, School of Applied Science, Cranfield University (2011)
10. M. Aien, A. Hajebrahimi, M. Fotuhi-Firuzabad, A comprehensive review on uncertainty modeling techniques in power system studies. Renew. Sustain. Energy Rev. **57**, 1077–1089 (2016)
11. H.-S. Hwang, Costing RAM design and test analysis model for production facility. Int. J. Prod. Econ. **98**, 143–149 (2005)

CFD Analysis of Fouling Effects on Aerodynamics Performance of Turbine Blades

A. T. Baheta, K. P. Leong, Shaharin Anwar Sulaiman and A. D. Fentaye

Fouling on gas turbine blades is detrimental to process operation as it may, over a period of time, reduce the blade efficiency and consequently the turbine's efficiency. With the limitation of today's technology, experimental study or real-life observation of fouling in a gas turbine is beyond the imagination of maintenance engineers. Hence, the effect of fouling cannot be fully quantified for the engineers to come out with mitigation or intervention plans. Nevertheless, computational fluid dynamics (CFD) may provide a good simulation to understand the phenomena. In this chapter, a recent effort involving CFD study on the influence of fouling on the gas turbine performance is presented. Firstly, the nature of fouling on gas turbine and the general consequences are discussed. This is followed by an elaboration on how CFD study has been conducted by the authors. Finally, the findings from the study are discussed.

Introduction

Gas turbines are used in power generation systems especially in the oil and gas industries. Most offshore platforms depend on aero-derivative gas turbines to provide power. However, it is not an unusual practice for operators to lengthen the service hours for a turbine blade in order to save cost and time needed for periodic maintenance. When the service hours of a turbine are long, common degradation issues like fouling will occur. Fouling is the accumulation of air contaminants and combustion ashes on the blade surfaces. As the impingement of the hot flue gas continues over time, more and more particles lose kinetic energy and eventually land on top of the turbine blade due to friction and adhesive nature of the particles [1]. This

A. T. Baheta (✉) · K. P. Leong · S. A. Sulaiman · A. D. Fentaye
Universiti Teknologi PETRONAS, Seri Iskandar Perak, Malaysia
e-mail: aklilu.baheta@utp.edu.my

© The Author(s), under exclusive licence to Springer Nature Singapore Pte Ltd. 2019   73
S. A. Sulaiman (ed.), *Rotating Machineries*, SpringerBriefs in Applied Sciences
and Technology, https://doi.org/10.1007/978-981-13-2357-7_5

phenomenon, in the end, causes thickening of the blade and affects the performance of the turbine blade.

The working environment of gas turbines at offshore platforms is harsh. Some of the common airborne contaminants are salt, sand, oil droplets, water mists, and soot [2]. As service hour increases, the level of the material deposit will also increase. As deposits accumulate on the surface, the contour of the blade surface becomes rough which leads to an increase in surface roughness. In addition, it also reduces the flow capacity between the blades. Other than that, since the accumulated particles thicken over time, this would cause an increase in the overall temperature of the turbine blades [3].

With this in mind, an analysis is required to study the factors which contribute to fouling as well as how fouling affects the performance of the blade in terms of temperature and pressure distribution. In this chapter, the factors that can cause fouling and the effect of fouling on the aero-thermodynamic properties of the gas turbine blades are investigated. Moreover, the most affected region of the blade is identified.

Gas Turbine Fouling and Its Effects

Gas turbine consists of different components including a compressor, combustion chamber, and turbine. The most widely used gas power cycle is Brayton Cycle which comprises four stages, namely isentropic compression in the compressor, isobaric heat addition in the combustion chamber, isentropic expansion in the turbine, and lastly isobaric heat rejection process, which cools and circulates the hot exhaust gases for the next cycle in an ideal working environment [4]. When the ambient air and the hot gas come in contact with the compressor and turbine blades, this would lead to common gas turbine fouling and erosion. Salt, sand, oil mist, smoke, and carbon are some of the air contaminants that can cause fouling [5]. As service hours increase, particles tend to be deposited on the suction surface and leading edge as compared to the pressure surface and trailing edge of the turbine blades [6]. In this case, erosion is said to be the abrasive removal of material which is caused by the hard particles larger than 10 μm hitting on the turbine blades [7, 8]. With that, it is critical to differentiate fouling from erosion as both of them have a different impact on the well-being of turbine blades.

One of the most significant properties which differentiate erosion from fouling is that erosion causes pitting or decaying action on the surface whereas fouling results in thickening of the surface due to the accumulation of particles. It is believed that particles that are bigger than 10 μm will cause fouling whereas smaller particles of smaller size contribute more towards erosion. For example, it has been approximated that for a Frame 9351FA gas turbine which has an ISO flow rate of 1429 lb/s, it would take in about 225 tons of foulant a year at 10 ppm foulant loading rate [3].

Figure 1 shows a simplified temporal variation of deposit thickness for a gas turbine, as adapted from Dalili et al. [9]. Generally, as flue gas travels from nozzles

Fig. 1 Fouling process over time adopted

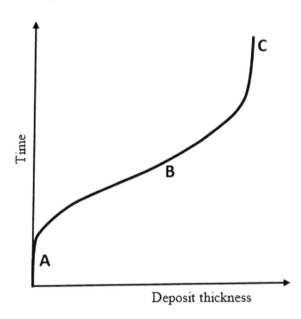

with high speed and impinges on the surface of the turbine blades, the initiation of the adhesion process begins at Point A in Fig. 1, whereby foreign particles are starting to accumulate. As the accumulation continues and reaches a steady state, the thickening of the blade profile will also become consistent over time in Stage B. As time passes, particles accumulate on the blade surface causing an increase in the weight of the blade, and at the same time, removal of particles happens due to increasing momentum caused by the weight of the particle deposit. This may result in an imbalance in the weight distribution. As the accumulation and removal of deposit occur simultaneously, deposit thickness is seen to increase gradually. While the deposit's weight increases, a high amount of deposit drop-off results in the removal process to dominate the accumulation process. Hence, the deposit thickness eventually reaches a constant rate as indicated at Point C [10]. The end result of fouling is deviations in flow capacity, pressure ratio, and isentropic efficiency [3]. Moreover, small colliding particles of less than 10 μm thicken and, in some severe cases due to the increased weight, cause vibration problems [11].

In the event that fouling occurs with an increase in operating hours, the flow capacity is seen to decrease as a result of particles deposition as indicated in Fig. 2. With that, less air is able to travel through the flow area and thus reducing the overall flow rate through the turbine.

Meher-Homji et al. [6] mentioned that surface roughness cannot uniquely characterize fouling phenomenon due to other degradation parameters that simultaneously happen on the surface like erosion, object damage, wear and corrosion, which complicate the process of studying fouling as an individual factor. However, Veer et al. [12] stated that the an accurate fouling deposite measurement is easy and straight

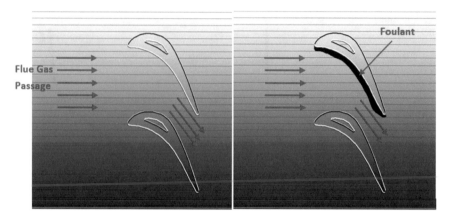

Fig. 2 Reduced flow area after fouling

forward based on time and age dependancy factors. Nonetheless, fouling was studied in this work based on its effect through surface area roughness and not through fouling factor itself due to CFD software limitations.

CFD Analysis Technique

In this study, first, information on the surface of the turbine blades is collected and analysed especially in terms of surface conditions. This helps to accumulate data for simulation using ANSYS Fluent software to replicate the surface conditions of the blades. One of the many ways to gather accurate surface conditions is to put a turbine blade that has been in operation for a long period under the microscope. By doing so, an estimate of the surface roughness can be obtained without complete dependencies on the simulated and assigned version of surface area roughness in ANSYS Fluent.

In the first place, 3D scanning is applied in determining the geometry and characteristics of the turbine blades. After performing the transfer of node points to GEOMAGIC and removal of faults, modelling is done using CATIA V5 software. The actual turbine blades are shown in Fig. 3a and the corresponding models that are generated in CATIA are displayed in Fig. 3b. Following this, in order to develop the blade model for CFD analysis, ANSYS Fluent is utilized. For this purpose, surface roughness characteristics data are necessary.

Surface area roughness is defined as the deviated contour of a real surface from its ideal form as caused by the accumulation of particles. It can be measured using different ways such as the root-mean-square method:

$$\text{Root Mean Square} = \sqrt{\frac{1}{n}\sum_{i=1}^{n} y_i^2} \tag{1}$$

(a)   **(b)**

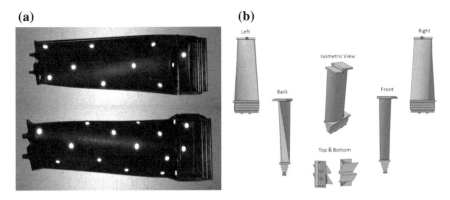

Fig. 3 Turbine blade samples: **a** for 3D scanning, **b** CATPART model

where y is the vertical deviation.

The data on blade surface conditions are mimicked through assignment of a numerical value in ANSYS Fluent on the design of the turbine blade mesh model. The programmed value in which ANSYS Fluent accepted is in the range of zero to one with 0.5 being a smooth surface with absolute zero deviations on the surface. Factor 1.0 in this case implies a rough surface with positive deviations on the surface, while the zero point represents a surface with negative deviations on the surface. In this study, only fouling factors from 0.5 to 1.0 were considered since the presence of fouling indicates positive deviations on the turbine blade surface.

Once the collection of surface roughness data has been completed, the construction of mesh model in ANSYS Fluent software can then be performed. In order to eliminate any sort of discrepancies, the generated mesh model was replicated based on an existing gas turbine design like the one displayed in Fig. 3. By that, the mesh model will have to adhere to strict design codes as found in American Petroleum Institute (API) 616 and American Society of Mechanical Engineers (ASME) Performance Test Code (PTC) 22.

Once the mesh modelling process is completed, simulation of various parameters such as airflow, pressure, and temperature distribution is performed using CFD. The predicted values are then served as a benchmark signifying the possible phenomena that occurred in a real-life situation. The complete process flow of the proposed investigation technique is presented in Fig. 4.

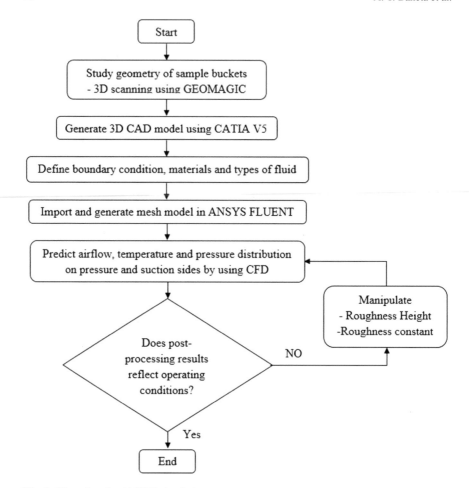

Fig. 4 Flow chart for ANSYS simulation

Findings

Fouling Effects on Airflow Across Blade Surface

The uneven surface caused by fouling is believed to have an imminent effect on the airflow across the blade span. Initial simulation using ANSYS Fluent shows that fouling causes substantial amount of turbulence during flow separation on suction side as displayed in Fig. 5.

The simulation was conducted using foulant particles with the size of 450 μm. Although the particles are very small, high amount of turbulence could be seen at the separation of the flow and this eventually caused an increase in drag due to the additional mix of air from the boundary layer to the air around it.

Fig. 5 Airflow comparison
for smooth and rough surface

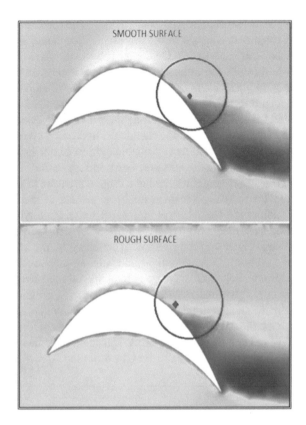

Drag, in this case, happens due to the surface roughness and is known as skin friction drag [13]. Similar events can also be found on the disrupted airflow due to ice accretion on the airfoils of an airplane. The red point in Fig. 5 indicates the separation point for both scenarios. It can be seen that the point shifts upward, towards the tip of the suction side, when large-sized foulant is introduced.

During this process, the shifting of the separation point causes the formation of eddies which are visible in the blue waves-like structure on the right side of the blade. As mentioned by Schultz and Swain [13], the size of these eddies is a good indicator of the magnitude of the wake and eventually the turbulence of the flow. The simulation results, as shown in Fig. 7, indicate that the area affected the most by fouling is at the base of the blade (leading age of the blade).

Fouling Effects on Pressure Distribution Across Blade Surface

One of the objectives of this study is to find out the relationship between surface roughness and pressure distribution. During simulation, two parameters were used to replicate the roughness intensity, namely wall roughness and wall roughness constant. A roughness constant of 0.5 represents the smooth surface, and it increases up to a maximum value of 1.0. On the other hand, particles of 0, 150, and 450 μm are used to investigate the influence of surface roughness on the pressure distribution curve at 50% blade span. The 0 μm size is included as part of the boundary conditions, which resembles a new and clean blade to be used as a benchmark to signify the effect of an increase in wall roughness. Dust, oil leaks, sand, insects, ash, salt, smog, aerosols, clouds, fog, ice, chemicals, hydrocarbons, and moistures are some of the gas turbine contaminants available in most of the application environments [14].

The simulation results are shown in Figs. 6 and 7. One can see that the increase in roughness height resulted in a drop in pressure throughout the blade surface. The effect is more significant at the lower surface of the blade especially at the middle region of the chord as this area experiences the highest airflow. An increase in the roughness height from 0 to 150 μm saw a decrease by about 90 kPa, which makes up about 15% of the total pressure experienced by the lower surface blade at the center of the chord. In roughly the same vein, an increase from 150 to 450 μm results in approximately 10% decrease in total pressure located at the mid-section of the lower surface of the blade. With that, the change in pressure is seen to be more obvious with the smaller number of roughness height which, in this case, is 150 μm. In other words, the decrease of pressure distribution is not linear with respect to the increase in roughness height. What this graph suggests is that when fouling happens, the lift induced is reduced as a result of the differential pressure present in between pressure and suction sides.

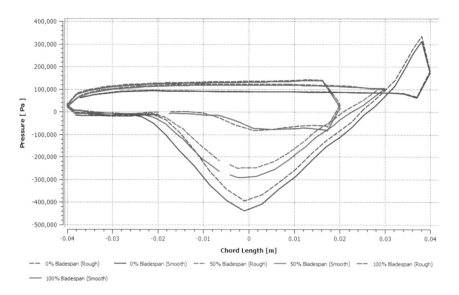

Fig. 6 Fouling effect on base end of bucket (0% blade span)

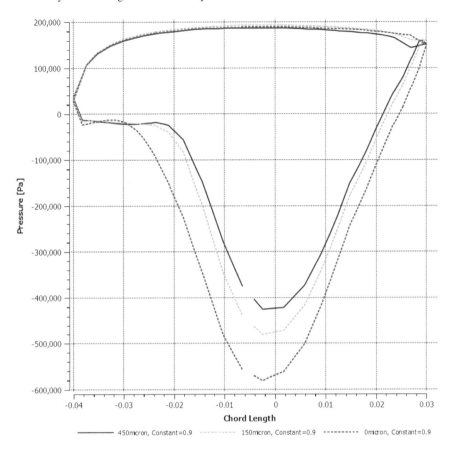

Fig. 7 Pressure distribution for different roughness heights (50% blade span)

When the simulation process is completed, the processing of the results is con-ducted using Fluent pre-processor, which is capable of producing a graphical rep-resentation in the form of contour shades and streamlines representing the airflow pattern across the blade. It is clearly shown that the lower end of the blade nearer to the base of the blade experiences a higher amount of pressure; see Fig. 8. In the figure, the yellow region on the pressure side of the turbine blade showed higher pressure value than the blue region on the suction side. This difference in pressure is the source of the force that leads to rotation of the rotary blades. This is in line with the basic operating principle of a turbine blade similar to an airfoil whereby the difference in pressure distribution causes lift and eventually the turbine blade, in this case, to rotate when connected to a shaft.

However, reduction in pressure distribution has been seen consistently throughout the pressure side of the blade. If the total area of reduction is taken into account, the base of the bucket exhibited the most decreased amount of pressure loss. As compared to the mid-region of the blade which is highlighted in purple lines, the base region

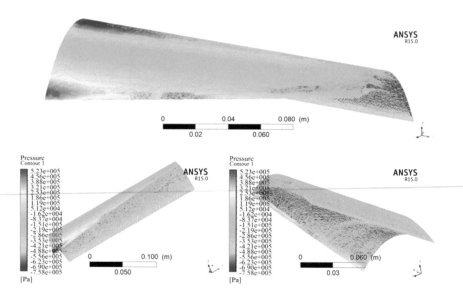

Fig. 8 Pressure distribution across blade surface (roughness constant factor = 1.0)

seemed to experience more reduction due to its chord length which spanned from −0.04 to 0.04 m whereas the mid-region only stretched from −0.04 to 0.02 m.

Fouling Effects on Temperature Distribution

In order to have a completed information about the effect of fouling on thermodynamic properties, the temperature distribution along the cord of the blade has been analysed. Figure 9 illustrates the temperature distribution of the blade at 50% blade span under different blade surface roughness factors. It is shown that as roughness constant increases the temperature distribution across the pressure side on the leading edge decreases. The temperature gets increasing as the roughness increases. This is because as roughness increases the friction between the flowing gas and the blade surfaces increases.

Summary

The main objective of this study is to investigate the effect of fouling on the aerothermodynamics properties of a gas turbine blade. For this purpose, a simulation model is developed utilizing ANSYS Fluent and CFD software. Then, the effect of surface roughness on the blade is investigated by introducing different contaminants of dif-

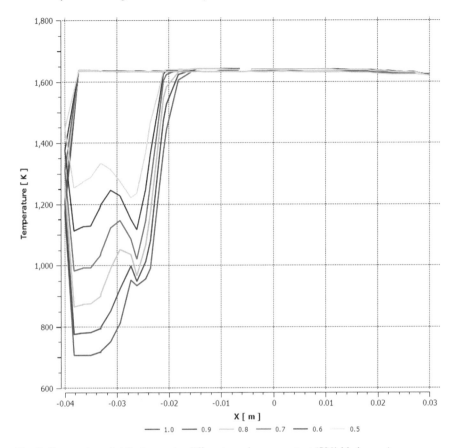

Fig. 9 Temperature distribution under different roughness constant (50% blade span)

ferent sizes. The simulation results for the temperature distribution across the blades reveal that temperature increases gradually as the level of fouling deposit increases. The increase in temperature is maximum on the base end of the blade span as that particular region has the highest surface area exposed to fouling. This leads to a loss in isentropic efficiency of the gas turbine components.

On the other hand, the simulation results for pressure distribution show that pressure decreases over the blade surface area as the severity of surface roughness increases. The base of the bucket experienced the most pressure loss as compared to the mid and tip region of the bucket. The pressure reduction ranges between 10 and 15% for a foulant with a size of 450 μm. A decrease in such a large amount of pressure on the pressure side results in a lower overall driving force.

Other than that, the existence of fouling on the gas turbine blades results in decrease in flow capacity of the engine. This would lead to a loss in overall efficiency and reduced power production.

References

1. A. Razak, *Industrial Gas Turbines: Performance and Operability* (Elsevier, Cambridge, 2007)
2. I.S. Diakunchak, Performance deterioration in industrial gas turbines. J. Eng. Gas Turbines Power **114**, 161–168 (1992)
3. C.B. Meher-Homji, A. Bromley, Gas turbine axial compressor fouling and washing, in *33rd Turbomachinery Symposium*, Houston, TX, Sept 2004, pp. 20–23
4. Y.A. Cengel, M.A. Boles, Thermodynamics: an engineering approach. Sea **1000**, 8862 (2002)
5. R. Kurz, K. Brun, Degradation of gas turbine performance in natural gas service. J. Nat. Gas Sci. Eng. **1**, 95–102 (2009)
6. C.B. Meher-Homji, M. Chaker, A.F. Bromley, The fouling of axial flow compressors: causes, effects, susceptibility, and sensitivity, in *ASME Turbo Expo 2009: Power for Land, Sea, and Air*, 2009, pp. 571–590
7. R. Kurz, K. Brun, Gas turbine compressor blade fouling mechanisms. Pipeline Gas J. **238**, 18–21 (2011)
8. A.A. Hamed, W. Tabakoff, R.B. Rivir, K. Das, P. Arora, Turbine blade surface deterioration by erosion. J. Turbomach. **127**, 445–452 (2005)
9. N. Dalili, A. Edrisy, R. Carriveau, A review of surface engineering issues critical to wind turbine performance. Renew. Sustain. Energy Rev. **13**, 428–438 (2009)
10. T.R. Bott, *Fouling of Heat Exchangers* (Elsevier, Amsterdam, 1995)
11. D. Brandt, R. Wesorick, *GE Gas Turbine Design Philosophy*, GER-3434, General Electric (1994)
12. T. Veer, K.K. Haglerod, O. Bolland, Measured data correction for improved fouling and degradation analysis of offshore gas turbines, 2004, pp. 823–830
13. M.P. Schultz, G.W. Swain, The influence of biofilms on skin friction drag. Biofouling **15**, 129–139 (2000)
14. M. Wilcox, R. Baldwin, A. Garcia-Hernandez, K. Brun, *Guideline for Gas Turbine Inlet Air Filtration Systems* (Gas Machinery Research Council, Dallas, TX, 2010)

Electrohydrodynamic Pumps for Dielectric Liquid Application

Mohammad Shakir Nasif

Operation and maintenance is a costly process in rotating equipment. Reducing this operation and maintenance is of significant economical and operational benefits in plants. Electrohydrodynamic pumping utilizes electric field to generate forces on dielectric liquids. These forces will push the liquids resulting in pumping effect. The pumping effect will be generated without the need of moving parts within the pump, hence significantly minimizing failure and required maintenance.

Introduction

In the last few decades, energy conservation demands and concern over global warning effects have been significantly increased. Rotating machines operation consumes significant energy, and its maintenance is costly. Hence, researchers focused their research on the development of equipment that is able to transport fluid with almost no energy consumption and with no moving parts.

Electrohydrodynamics (EHD) deal with the interaction between electric fields and fluid flow. This interaction can result in electrically induced pumping. EHD method utilizes the effect of electric field to induce a secondary motion in a dielectric liquid.

The amount of power used in the electric field is very small because of using dielectric liquids in comparison with the rate of pumping that in fact is the whole reason for being EHD is an interesting phenomenon in pumping, whereas the operating cost in terms of energy and money are very small. The major advantages of using EHD pumping are:

- Rapid and smart control of pumping rate by varying the applied electric field
- Non-mechanical and parts will not wear out

M. S. Nasif (✉)
Universiti Teknologi PETRONAS, Perak, Malaysia
e-mail: mohammad.nasif@utp.edu.my

- Applicable to single- and multi-phase flows
- Minimal power consumption
- Very low vibration and noise

Electrohydrodynamic Forces

Electrohydrodynamic (EHD) pumps utilize electrostatic forces acting on dielectric liquids to generate flow. There are several types of EHD pumps, and the distinction is based mainly on the method by which the charged particles are introduced into the fluid.

The body force acting on the fluid resulting from the interaction of a non-homogeneous electric field E with a fluid space charge density q is given by the relation [1]:

$$f = q\vec{E} - \frac{1}{2}E^2\nabla\varepsilon + \frac{1}{2}\nabla\left[E^2\left(\frac{\partial\varepsilon}{\partial\rho}\right)_T\rho\right] \tag{1}$$

where q, E and ε are the charge density, electric field and liquid permittivity, respectively; ρ is the mass density, and $(\partial\varepsilon/\partial\rho)_T$ is determined at constant temperature T.

The first term on the right-hand side of Eq. (1) is the Coulomb force, and it is called the electrophoretic force, which is the force acting on free charges and it will move along the electric field lines towards the opposite charged electrode. The total motion of the charged body depends on the charge relaxation time (τ) which is defined as the time, which the influence of the electric field will take place and it can be presented as [2]:

$$\tau = \frac{\varepsilon}{\sigma} \tag{2}$$

the second term is referred to as the dielectric force, and the third term is the electrostriction force. For single-phase, incompressible, isothermal dielectric liquids, the predominant mechanism for EHD is the Coulomb force.

There are several types of EHD pumps, and the distinction is based mainly on the method by which the charged particles are introduced into the fluid. The types are presented in the following sections.

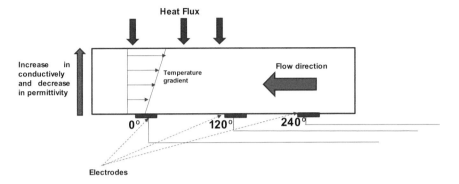

Fig. 1 A typical induction pump

Types of EHD Pumps

Induction Type

Induction-type EHD pumps require a gradient of fluid electrical conductivity or permittivity. This is typically achieved by fluid heating or by discontinuities in properties.

Alternating voltages are imposed on the electrodes present on the boundary of the fluid channel. These voltages vary in time, creating a travelling wave that moves through the working fluid, perpendicular to the gradient in electric conductivity as shown in Fig. 1.

The EHD induction pumping relies on the generation of induced charges. This charge induction in the presence of an electric field takes place due to non-uniformity in the electrical conductivity of the fluid which can be caused by the non-uniform temperature distribution. Therefore, this pumping mechanism cannot be utilized in an isothermal liquid.

Charges are induced by the travelling electric field waves at the interfaces of the media, or in the bulk of the working fluid where the gradients in conductivity or permittivity occur. The charges are attracted or repelled by the space- and time-varying electric field and carry with them the bulk fluid due to viscous effects. In a time period, charges neutralize on the order of the charge relaxation time. Hence, short distances between electrodes allow the charges to move from one electrode to the other before being neutralized [3].

The direction of motion is dependent upon the direction of the travelling wave and the temperature gradient. Induction EHD pumps are dependent upon the electrical properties of the fluid (permittivity and conductivity). Depending on the electric conductivity gradient generated of the pumped fluid, induction pumps can generate flow velocities ranged 10–80 μm/s [3]. The induction EHD pumps can be fabricated

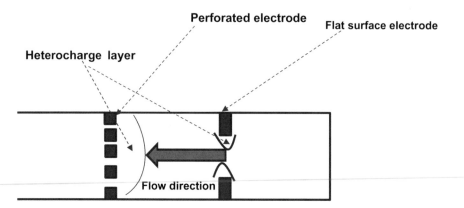

Fig. 2 Conduction-type pump

by using micromachining methods directly in the channel and hence enhancing ease of fabrication.

Conduction Type

In this type of pumping, there is not direct charge injection. The operating principle of this pump is based on the dissociation and recombination of neutral particles and asymmetric of the electric field [4–6]. The fluid direction is always from the narrower electrode, where the electric field lines are more packed together, to the wider electrode, where the electric field lines are less packed and weaker. In a neutral fluid medium, the rate of dissociation and recombination is in equilibrium. However, the intensity of the electric field in the vicinity of electrodes disturbs this equilibrium and forms a layer in which the neutral species are no longer at equilibrium. Due to the increase of dissociation rate compared to the recombination, free charges will be generated adjacent to the electrode and form a layer called heterocharge layer causing the fluid to be moved [4–6], as illustrated in Fig. 2.

Polarization Type

This micropump utilizes a non-homogeneous electric field through the fluid to create electric density variation. A gradient in the energy of the dipoles (from the liquid) is generated. Dipoles in regions of high electric field have lower energy. Hence, the fluid in the flow channel bounded by the electrodes will have lower energies as compared with the electrode region causing the fluid external to the electrode section to move into the flow channel causing resulting in pumping action [3], as shown in Fig. 3.

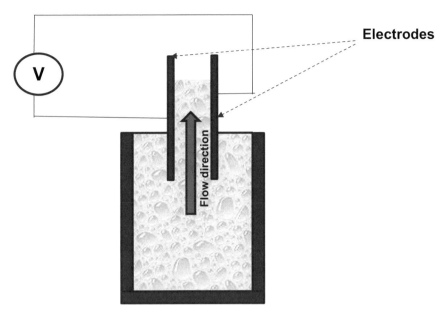

Fig. 3 Polarization-type pump

Ion-drag Type

The operating principle of ion-drag pumping is based on establishing a free charge which will be generated between a sharp-edged electrode called emitter and smooth-edged surface which is called collector. The two objects will be connected to high-voltage power supply in which the fluid is located between the emitter and collector.

When the electric field between the emitter and collector electrodes exceeds a certain threshold, free charges (ions) are injected to the fluid which has neutral molecules. As a result, the free charges move along the electric field lines (between the emitter and collector) and their motion cause collision with the fluid free charges resulting in a drag force along the channel. By increasing the voltage of electric field, the drag forces increase and eventually giving a rise to the fluid causing the fluid to move, as depicted in Fig. 4.

Pumping velocities as high as 33 cm/s were achieved with this type of pumps. Various hydrocarbon-based dielectric fluids were studied by researchers, and the results showed that pumping performance depended on fluid properties, primarily fluid viscosity and electrical conductivity [5].

Fig. 4 Ion-drag pump

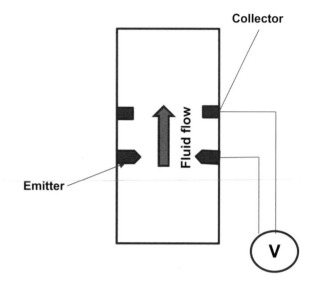

EHD Pumps Application and Performance

Due to different operating principles of EHD pumps, the performance and application of each pump vary. Details on the possible application and performance are explained in the following subsections.

Conduction EHD Pump

Researchers found that this pump can achieve maximum flow velocity of 8.9 cm/s and pressure head of 586 Pa at 20 kV with one pair of the multi-tube electrode design. A higher pressure head can be achieved by increasing the number of electrode pairs. The main application may be the pumping of refrigerant for chip-integrated cooling systems [5].

Ion-drag Pump

A potential application of an ion-drag pumping system is in pumping fluids in cryogenic cooling microsystems. It can also be used in fuel injection loops, and gas and liquid pumping where small quantities of dielectric fluids need to be pumped [7].

Table 1 EHD pumps' performance

Reference	Type	Pumped liquid	ΔP_{max} (Pa)	Q_{max} (mm³/s)
[8]	Ion-drag	HFE—7100	500	75
[9]	Ion-drag	Ethanol	2500	200
[10]	Ion-drag	Ethyl alcohol	175	0.25
[11]	Conduction	R 123	270	4000
[12]	Induction	Water	6	7.5×10^{-3}
[13]	Induction	Water	2.7	3×10^{-4}

Induction Pump

Although the generated pressure from this type of pumps is low, there are many systems where high pressure is not required such as generating flow mixing. This pump also can be utilized in macroscopic power generator to feed the microscopic device. Although the generated pressure is typically low, many applications do not require high pressure like the generation of secondary flows for mixing [7]. Table 1 shows the performance of each EHD pump based on the research performed.

Summary

This chapter presents the utilization of EHD phenomenon for pumping of dielectric liquids. In general, EHD pumping is a promising technique which requires no moving parts and consequently minimum maintenance. The power required to generate the pumping effect is negligible although high voltage is required. This is due to the very low current used which consequently result in low power consumption. This technique is still under research for larger scale; however, it is suitable for cryogenic applications and as well as cooling electronic of devices. Among four different types of EHD pumping, induction pumping is more associated with cooling operations due to its particular features. In comparison with the other three types of EHD pumping, ion-drag pumping has the potential to produce the highest flow rate although it requires direct injection of ions in the fluid. EHD conduction generates lower flow rate. However, conduction and polarization pumping applications require much more researches and investigations in order to optimize its performance and generated flow rate.

References

1. J.R. Melcher, Traveling-wave induced electro-convection. Phys. Fluids **9**, 1548–1555 (1966)
2. J. Ogata, Y. Iwafuji, Y. Shimada, T. Yamazaki, Boiling heat transfer enhancement in tube-bundle evaporators utilizing electric field effects. ASHRAE Trans. Symp. 435–444 (1992)
3. B. Iverson, S.V. Garimella, Recent advances in microscale pumping technologies: a review and evaluation. Microfluid. Nanofluid. **5**, 145–174 (2008)
4. K. Aryana, A. Ghiami, M. Edalatpour, M. Passandideh-Fard, A review on electrohydrodynamic (EHD) pumps, in *24th Annual International Conference on Mechanical Engineering-ISME2* (Yazd University, Yazd, Iran, 2016)
5. J. Seyed-Yagoobi, Electrohydrodynamic pumping of dielectric liquids. J. Electrostat. **63**, 861–869 (2005)
6. S. Jeong, J. Seyed-Yagoobi, Experimental study of electrohydrodynamic pumping through conduction phenomenon. J. Electrostat. **56**, 123–133 (2002)
7. A. Ramos, Electrohydrodynamic and magnetohydrodynamic micropumps, Chap. 2, in *Microfluidic Technologies for Miniaturized Analysis Systems* (Springer, USA, 2007), pp. 59–116
8. J. Darabi, H. Wang, Development of an electrohydrodynamic injection micropump and its potential application in pumping fluids in cryogenic cooling systems. J. Microelectromech. Syst. **14**, 747–755 (2005)
9. A. Richter, H. Sandmaier, An electrohydrodynamic micropump, in *Proceedings of Micro Electro Mechanical Systems, An Investigation of Microfluidic Technologies for Miniaturized Analysis Systems*, 1990
10. S.H. Ahn, Y.K. Kim, Fabrication and experiment of planar micro ion drag pump. Int. Conf. Solid State Sens. Actuators Transducers **1**, 373–376 (1997)
11. Y. Feng, J. Seyed-Yagoobi, Understanding of electrohydrodynamic conduction pumping phenomenon. Phys. Fluids **16**, 2432–2441 (2004)
12. G. Fuhr, T. Schnelle, B. Wagner, Travelling wave-driven microfabricated electrohydrodynamic pumps for liquids. J. Microelectromech. Syst. **4**, 217–226 (1994)
13. M. Felten, P. Geggier, M. Jager, C. Duschl, Controlling electrohydrodynamic pumping in microchannels through defined temperature fields. Phys. Fluids **18**, 051707 (2006)

Experimental Study on Electrical Power Generation from a 1-kW Engine Using Simulated Biogas Fuel

Suhaimi Hassan and Hamdan Ya

This chapter presents a simulated biogas with mixture of 60% methane (CH_4) and 40% carbon dioxide (CO_2) as a fuel for small 1 kW spark-ignition engine (SI). The simulated biogas fuel was tested on SI engine, and the engine performance was investigated at the constant engine speed under load transients. The SI engine was connected to load bank to generate a comparable result by using gasoline, liquefied petroleum gas (LPG), and natural gas as fuel in the engine to be evaluated with simulated biogas. The result showed that maximum specific fuel consumption for LPG and natural gas was decreased by 10 and 23.4%, respectively, when compared to gasoline, and it is proven that the simulated biogas has consumed more fuel (1254.83 kg/kWh) with only reach up to 780 W. The power reduction of engine using simulated biogas was about 22% as compared to gasoline. In terms of engine efficiency, gasoline, LPG, and natural gas have generated 21, 20.7, 20.4%, respectively, while simulated biogas generated only 0.16% of engine efficiency.

Introduction

The primary energy source such as crude oil, natural gas, and other conventional fuels are inadequate resources formed by geological processes throughout solar energy buildup into the earth over millions of years. Furthermore, the fossil fuels will significantly contribute to the emission of greenhouse gases (GHG) from the combustion and raising the climate change issue. In regard to environmental point of view, there is an urge in reducing the emission of pollutant substances in the atmosphere. According to Pusat Tenaga Malaysia (PTM), biogas is among several renewable sources of energy that will be prioritized under the policy [1]. The conversion of biogas energy

S. Hassan (✉) · H. Ya
Universiti Teknologi PETRONAS, 32610 Bandar Seri Iskandar, Perak, Malaysia
e-mail: suhaimiha@utp.edu.my

© The Author(s), under exclusive licence to Springer Nature Singapore Pte Ltd. 2019
S. A. Sulaiman (ed.), *Rotating Machineries*, SpringerBriefs in Applied Sciences and Technology, https://doi.org/10.1007/978-981-13-2357-7_7

is presented as a solution to the large volume of waste produced, as it allows the reduction of the toxic potential of CH_4 emissions [2].

The biogas is obtained by the means of anaerobic digestion where the fermentative process without oxygen (O_2) is take place. For this process to occur, the degradation of organic matters in landfills, effluent treatment plants, and animal waste happened [2]. Different researchers reported that with various compositions [3], proved that the concentrations of the biogas composition are dependent on the substrate composition from which the gas was produced. Furthermore, biogas is predominantly important because of likelihood of use in internal combustion engines compared to others, which are the main power source for transport vehicles and also commonly used for powering of generators of electrical energy. This possibility of use is justified by biogas properties, which make it suitable for internal combustion engines, ICE [4]. Basically, the classification of engine of is based on type fuel used. Gasoline is one of volatile liquids, whereas biogas and LPG are categorized as gaseous fuels. It is said that engine using gaseous fuels has similar working principle as the engine using volatile liquids [5].

Usually, the engine test results obtained in terms of power output, specific gas consumption, and thermal efficiency [3]. Various high-efficiency strategies for power generation using biogas and the results were compared with gasoline, LPG, and natural gas operation at same electrical power. Several researchers [6, 7] have studied on the effect on CO_2 in biogas-fueled engines, and they found that the reduction in concentration of CO_2 leads to higher efficiency and power output in SI engine. Additionally, in regard to increase efficiency of biogas-fueled generator, Surata et al. [3] stated that the LPG was added to the mixture up to 80% biogas and 20% LPG. Moreover, in their study, the engine test results of enriched biogas containing 95% of CH_4 has showed that engine performance is comparable to compressed natural gas which proved that biogas can be used as fuel for natural gas vehicles [3, 8]. Other researcher [9] mentioned that the spark-ignition engine operation with biogas containing significant fractions of inert gases such as carbon dioxide (CO_2) and nitrogen (N_2) exhibits penalties of performance compared with natural gas or gasoline. The typical properties of biogas compared to other gaseous fuels are shown in Table 1 [10].

Methodology

The experimental work was conducted to evaluate and examine the direct use of gasoline, LPG, natural gas, and biogas in a 1-kW spark-ignition portable engine in terms of the engine performance and exhaust emissions at different electrical load conditions. Engine power evaluation was made by comparing the output capacity of single cylinder four-stroke SI engine as specifications are shown in Table 2.

In this study, the load bank was used to measure electric power produced by the electric generator. It consists of ten equal lightbulbs of 100 W each, wired in parallel to allow easy load variation for flexible testing. To measure the current of

Table 1 Fuel gas properties

Property	Gasoline	LPG	Natural gas	Simulated biogas
Lower heating value at 1 atm and 15 °C (MJ/kg)	42.9	45.7	50.0	17.64
Density at 1 atm and 15 °C (kg/m^3)	750	2.26	0.7–0.9	1.43
Flame speed (cm/s)	62	38.25	34	25
Stoichiometric A/F (kg of air/kg of fuel)	14.7	15.5	17.3	11
Auto-ignition temperature (°C)	246–280	405–450	540	625

Table 2 Specification of the engine

Model	EV10i
Engine type	4-stroke, OHV single cylinder
Displacement	49 cc (cubic centimeter)
Compression	7.5:1
Rated revolution	5500 r/min
Fuel tank capacity	0.61 Gallon (2.3 L)

engine electric generator consumed, an ammeter is used. Emission analysis was conducted with gas analyzer known Automotive Exhaust Gas Analyzer (AUTOplus 5-2). Meanwhile, temperature data was also taken using Portable Handheld Data Logger OM-DAQPRO-5300 model as it contains thermocouple temperature sensors.

Procedure Running the Generator with Gasoline

The generator is weighed using TCS-Z Series Electronic Weighing Platform Scale. The generator fueled with gasoline's weight is recorded. Start button is pushed, and conversion to Gasoline gate is ensured. The choke lever is turned on to close which is to the right side. Next, the grip starts and the rope are pulled to the distance of 1.0 m in 0.4 s at full tilt. The pulling procedure will need to repeat for 3–5 times for the cold engine. After the engine is started, the choke lever is turned to the left side to open.

After the engine is started and the output indicator (green) came out, the appliance is plugged in. Digital stopwatch is set to measure 15 min time taken for the experiment. On the first test to find out amount of current consumed for each bulb, the bulb

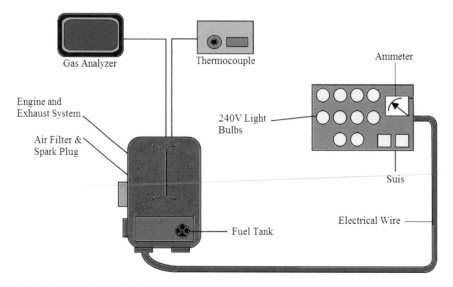

Fig. 1 Engine with load bank fueled with gasoline

is switched on one by one. On the second test, five bulbs are switched on for average load measurement and ten bulbs are switched on for maximum load measurement, respectively. The observation on the variation of the temperature with load as well as exhaust emission is recorded. The generator is weighed again for each test. The schematic diagram for running the generator with gasoline is shown in Fig. 1.

Procedure Running the Generator with LPG

LPG regulator is installed with three screws on the door to the plane window. The LPG trachea and regulating valve are fixed to make sure the various interfaces are locked on top of the LPG gas cylinder tank. The start switch is turned on. Conversion to brake rotation to gasoline gate is made. Valve on the LPG cylinder is opened. In order to start the engine, simultaneously the choke lever is pushed to the right side to close and the engine is started by pulling the grip starts and the rope to the distance of 1.0 m in 0.4 s at full tilt. Starter grip is pulled to operate the recoil starter to crank the engine. Then, the choke lever is turned to the left side to open, and brake rotation to LPG gate is converted. Two minutes is estimated for the conversion of the engine to completely finish up the small quantity of gasoline used when it had switched to gasoline gate in the first place. The knob on the LPG regulator is well-controlled to achieve stable operating condition at average and maximum load setting, respectively. The schematic diagram for running the generator with LPG is shown in Fig. 2.

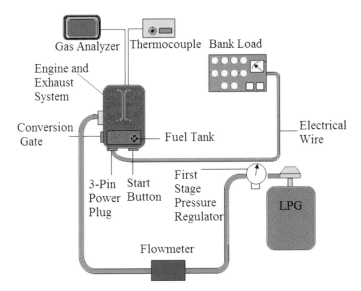

Fig. 2 Engine with load bank fueled with LPG

Procedure Running the Generator with Natural Gas

Single-stage regulator is installed on the natural gas inlet. The regulator is closed by turning the adjusting knob to the full counterclockwise position. The adjusting knob is turned clockwise. High-pressure hose is connected to the other side of regulator and connecting it to flowmeter before joined to the generator. The subsequent experimental setup is the same as running the generator with LPG to turn on the start switch. The schematic diagram for running the generator with natural gas is shown in Fig. 3.

Procedure Running the Generator with Simulated Biogas

The engine fuel system is modified by adding a CO_2 cylinder tank and two flow metering system for both CH_4 and CO_2 cylinders as in Fig. 4. Leak test is carried out at the tube connections. The engine is started. CH_4 and CO_2 flows are turned on, and the pressure regulator for both CH_4 and CO_2 is monitored for its percentage calibration. Another flowmeter is used to control the amount of gas mixture to support the load applied.

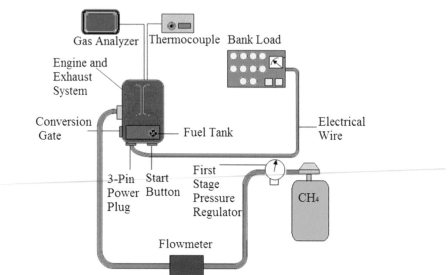

Fig. 3 Engine with load bank fueled with natural gas

Results and Discussions

Engine Performance

Figure 5 shows the variation of fuel flow rate with load. For LPG and natural gas, low fuel flow rate was needed to generate enough heat input to support the load applied. Furthermore, at average load, the fuel flow rate for LPG and natural gas was decreased to 5 and 11.5%, respectively. Meanwhile, at maximum load, the fuel flow rate was decreased to 10 and 23.4%, respectively, as compared to gasoline. Using simulated biogas, higher fuel flow rate was needed to generate enough heat input to support the load applied. As this engine has fixed engine speed, the air flow was limited. Therefore due to limited supply of gas and lower content of methane in the simulated biogas, the maximum load that the engine can take up is around 780 W as compared to gasoline, LPG and natural gas. Hence, the maximum load that the engine was capable of supporting gas decreased, and it was observed that it can only supply load up to 780 W due to lower methane content in the simulated biogas compared to natural gas.

Figure 6 shows the variation of specific fuel consumption with load. Compared to gasoline, running the engine on LPG resulted in around 10% increase in the fuel consumption to produce 1000 W of power.

Natural gas has the highest heating value where for the engine to develop the same power as when using gasoline, it already has satisfactory fuel and hence, it has lower specific fuel consumption (sfc), as compared to gasoline fuel. It resulted in

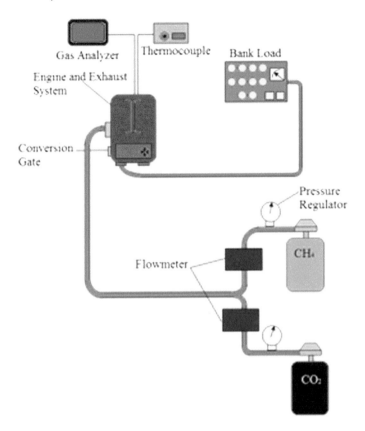

Fig. 4 Engine with load bank-fueled simulated biogas

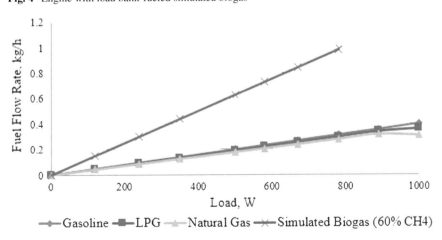

Fig. 5 Variation of fuel flow rate with load

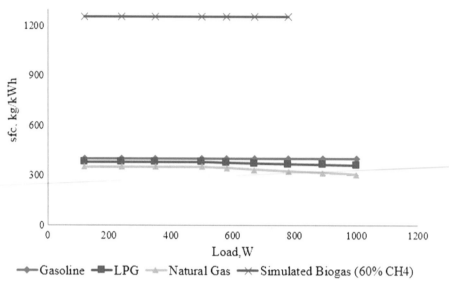

Fig. 6 Variation of specific fuel consumption with load

23.4% decreased in consumption to produce 1000 W when compared to gasoline. For running with simulated biogas, the values of specific fuel consumption were much higher compared to running with gasoline. Specific fuel consumption was 1254.83 kg/kWh up to 780 W only for simulated biogas with 60% methane. It resulted in 68.11% increase in consumption and power reduction of 22% when achieved 800 W load.

Figure 7 shows the variation of engine efficiency with load. Gasoline has the engine efficiency of 21% which was close to the theoretical highest value of engine efficiency. LPG, natural gas, and simulated biogas have generated 20.7, 20.4, and 0.16% of engine efficiency correspondingly. Calorific value of simulated biogas was lower in value than the other fuels though its mass flow rate was not much vary compared to gasoline caused it to have lower engine efficiency.

The variation of exhaust temperature with load is shown in Fig. 8. It was found that natural gas has the highest temperature as it contained purified methane that has different combustion features than simulated biogas because of percentage of CO_2 content. It combusted faster and at high temperature that required different adjustment of ignition timing. Even more, another characteristic of simulated biogas was that the temperature of its flame is high, where it proved that its exhaust gas temperature was in the high range.

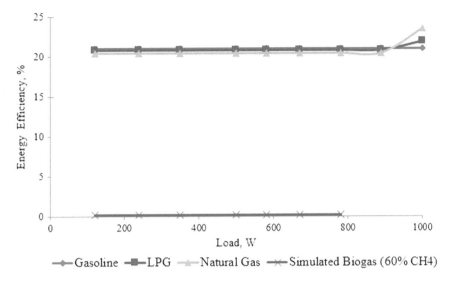

Fig. 7 Variation of engine efficiency with load

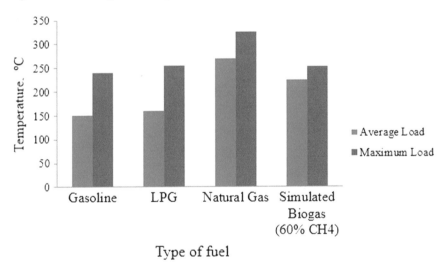

Fig. 8 Variation of exhaust temperature with load

Exhaust Emissions

Figure 9 illustrates the percentage of carbon monoxide (CO) in exhaust emission. CO emission was raised sharply for gasoline and simulated biogas as the reason it ran with the fuel-rich mixture where in the presence of CO revealed it has undergone incomplete combustion. Natural gas has resulted significantly lower emission of CO.

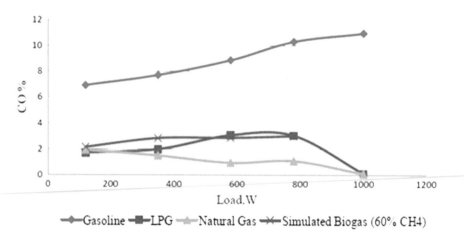

Fig. 9 Variation of CO emission with load

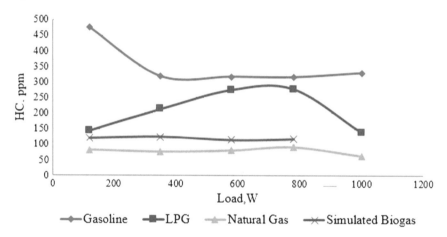

Fig. 10 Variation of HC emission with load

Figure 10 shows that hydrocarbon (HC) emission had similar tendencies to CO emissions. The HC emissions were increased with CO_2 percentages and decreased with electrical load. Also, between 0.6 and 0.8 kW load conditions for all fuels, the HC emissions are about to constant for all fuels as CO_2 blended steadily despite increase in load.

Figure 11 shows the percentage of nitrogen oxides (NO_X) at different load setting. NO_X emission was lower when the content of CO_2 in the fuel is in a high amount. However, NO_X increased with electrical load. NO_X formation was straightforwardly related to the flame temperature in an engine cylinder. The higher the flame temperature, the more the formation of NO_X.

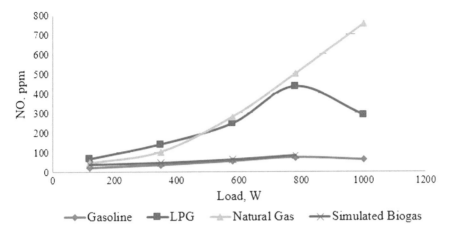

Fig. 11 Variation of NO_x emission with load

Summary

The study on the performance and emission of 1-kW SI engine has been successfully investigated with minimum modification on the engine to suit the usage of simulated biogas fuel. The study on the performance and emission of 1-kW SI engine has been successfully investigated. Generator with the minimum possible modifications. Engine performances such as fuel flow rate, specific fuel consumption, engine efficiency, and exhaust emissions were measured with various types of fuels and engine loads. Based on the emission analysis, natural gas represented a good fuel where it emitted lower emissions from exhaust system while simulated biogas also has brought significant reductions in CO, HC, and NO_X, and it is proven would help to reduce harmful greenhouse gas emission.

For future work, it is recommended to use actual biogas from nature sources as using simulated biogas has been satisfactory comparable with gasoline, LPG, and natural gas. In addition to the built-in automatic throttle (speed) control mechanism, the fixed speed engine needed additional flow regulation to control liquefied gas as well as compressed gas flow to support the variation in the electric load applied. Should be there is improvement of mixing chamber and cooling system of the engine. It is better if simulated biogas is already in the mixture form. Hence, the calibration on percentage of CH_4 and CO_2 from different cylinder tank could be neglected. From the result, the composition itself needs to change to a more suitable composition as to reduce the unnecessary emission from exhaust system while increasing its efficiency.

References

1. T.H. Oh, S.Y. Pang, S.C. Chua, Energy policy and alternative energy in Malaysia: issues and challenged for sustainable growth. Renew. Sustain. Energy Rev. **14**, 1241–1252 (2010)
2. V.P. Garcilasso, S.M.S.G. Velazquez, S.T. Coelho, L.S. Silva, Electric energy generation from landfill biogas: case study and barriers. In:*International Conference on Electrical and Control Engineering (ICECE)*, 16–18 Sept 2011
3. I.W. Surata, T.G.T. Nindhia, I.K.A. Atmika, D.N.K.P. Negara, I.W.E.P. Putra, Simple conversion method from gasoline to biogas fueled small engine to powered electric generator. Energy Procedia **52**, 626–632 (2014)
4. S. Mihic, Biogas fuel for internal combustion engines. Ann. Fac. Eng. Hunedoara, Tome II, Fascicole 3, pp. 179–190 (2004)
5. V. Ganesan,*Internal Combustion Engine*, 2nd edn. (McGraw-Hill, 2004)
6. E. Porpatham, A. Ramesh, B. Nagalingam, Investigation on the effect of concentration of methane in biogas when used as a fuel for a spark ignition engine. Fuel **87**, 1651–1659 (2008)
7. B.P. Pundir, I.C. Engines, *Combustion and Emissions* (Alpha Science International Ltd., Slough, United Kingdom, 2010)
8. M. Himabindu, R.V. Ravikrishna, Performance assessment of a small biogas-fuelled power generator prototype. J. Sci. Ind. Res. **73**, 781–785 (2014)
9. R.J. Crookes, Comparative bio-fuel performance in internal combustion engines. Biomass Bioenerg. **30**, 461–468 (2006)
10. EIA,*International Energy Outlook* (Energy Information Administration, Department of Energy, U.S.A, 2010)

Printed in the United States
By Bookmasters